Discrete and Continuous Simulation

Theory and Practice

Discrete and Continuous Simulation

Theory and Practice

Susmita Bandyopadhyay
Ranjan Bhattacharya

CRC Press
Taylor & Francis Group
Boca Raton London New York

CRC Press is an imprint of the
Taylor & Francis Group, an **informa** business

CRC Press
Taylor & Francis Group
6000 Broken Sound Parkway NW, Suite 300
Boca Raton, FL 33487-2742

First issued in paperback 2017

© 2014 by Taylor & Francis Group, LLC
CRC Press is an imprint of Taylor & Francis Group, an Informa business

No claim to original U.S. Government works
Version Date: 20140501

ISBN 13: 978-1-138-07699-0 (pbk)
ISBN 13: 978-1-4665-9639-9 (hbk)

Dedication

This book is dedicated to our parents.

Contents

List of Figures

List of Tables

Preface

We are in the age of globalization and are facing frequently changing global market environments in every sector. This dynamic environment enhances the level of complexity at significant levels. Identifying problems in such complex environments in any field of study is an extremely challenging task, especially if the time for finding a remedy to the problem is short. The search for solutions to problems in such situations becomes even more complicated and rigorous because of the rapidly increasing number of impatient customers who demand prompt solutions to their problems and demands. Under this scenario throughout the entire world, entrepreneurs and practitioners, especially in the fields of science and technology, need efficient tools to identify the problems inherent in their systems promptly so that they can visualize their problems to rectify them.

Simulation is such a method by which the problems inherent in a system can be identified with less effort and time. This can be accomplished with very small investment without affecting the original system as a whole. There is, of course, a risk related to the simulation experiment. This risk lies in selecting the right simulator and the right expert who can choose a simulator and can use it properly. However, if the simulator is selected properly (which is not a very difficult task), then simulation can be a very effective tool for the purpose. This explains the relevance of studying simulation. However, several universities and colleges throughout the entire world teach simulation to their students, which also highlights the importance of the subject.

This book shows various aspects of simulation study. The aspects start from continuous simulation and span over many other aspects including the simulation with system dynamics. The readers are expected to get a good overview of a variety of aspects of simulation through this book. We have exerted our sincere efforts to make this book a success. We expect that the readers will benefit from this book to a great extent.

Susmita Bandyopadhyay
Jadavpur University

Ranjan Bhattacharya
Jadavpur University

MATLAB® is a registered trademark of The MathWorks, Inc. For product information, please contact:

The MathWorks, Inc.
3 Apple Hill Drive
Natick, MA 01760-2098, USA
Tel: +1 508 647 7000
Fax: +1 508 647 7001
E-mail: info@mathworks.com
Web: www.mathworks.com

1 Introduction to Simulation

1.1 INTRODUCTION

Simulation can be defined as the imitation of a system through a prototype of the system to find the flaws and problems inherent in the system so as to rectify them. Thus, any kind of program imitating a system or process can be termed as "simulation." This chapter provides an introduction to the concept of simulation. Before describing the various aspects of simulation in this chapter, some of the definitions should be clarified. Thus, the concepts of system and model are described in this section since the words "system" and "model" are frequently used while studying simulation.

A system can be defined as a collection of interrelated components or units or entities. For example, a queuing system in a bank consists of several components such as customers, machines, clerks, managers, other employees, papers, and other things that are required for running the business. Similarly, the inventory system consists of items for inventory, maintenance clerk, the manager of the inventory control department, and customers who purchase the inventory. A retail store can be thought of as a system that consists of various products, customers who purchase the products, the counters, the customer service executives at the counters, other staff of the store, the manager of the store, the computers, and other machines and equipments. A waiting queue for treatment in a hospital can be thought of as a system that consists of doctors, nurses, patients, other associated staff, and equipments.

There are two types of systems—discrete and continuous. In a discrete system, the state variables' values change at discrete points in time, whereas in a continuous system, the state variables' values change continuously over time. The state variable of a system describes the state that reflects the values of the variables. Experiments are conducted to find the states of a defined system. Experimentation can be done on the actual system or a prototype of the system. If the actual system is large, it is difficult to observe the functioning of each of the components. On the contrary, developing a representative prototype for the system is a difficult task. If prototyping is not done properly, the purpose of simulation study will be in vain. There are two types of prototypes—physical and mathematical [1], also known as physical models and mathematical models, respectively. A physical model is a miniature of the actual system, whereas a mathematical model emphasizes only mathematics to represent the system on hand. Mathematical models can further be subdivided into two types: analytical systems, which use hardcore mathematical expressions to represent the working of a system, and simulation systems, which use mathematics but not in the form of explicit mathematical expressions. Simulation is entirely dependent on the uncertainties of a real system and thus uses probability distributions wherever

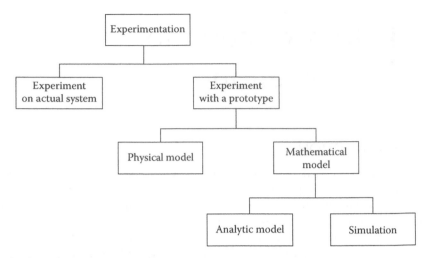

FIGURE 1.1   Type of experimentation.

the condition of uncertainty is applicable. Figure 1.1 shows the type of experimentation that can be conducted to study a system.

The figure clarifies the idea of a model. A model is a compact prototype of a system. It can be either a static model or a dynamic model. A static model represents a system at a particular point of time, whereas a dynamic model represents a system that changes with the advent of time. A model can also be a deterministic model or a stochastic model. A deterministic model has deterministic variables, that is, the variables that define the system are all deterministic in nature. The variables in a stochastic model are probabilistic in nature, that is, probability values or probability distribution can be used to depict the occurrences of the event represented by the variables. A model can be a discrete model or a continuous model. The variables in a discrete model change their values at discrete or individual points in time, whereas the variables in a continuous model evolve as the time advances.

The sections of this chapter are arranged as follows: Section 1.2 depicts the types of simulation. Section 1.3 describes the steps of simulation study. Section 1.4 mentions the areas of applications where the concept of simulation can be applied. Sections 1.5 and 1.6 provide examples of a queuing system and an inventory system, respectively. Section 1.7 enlists the advantages and disadvantages of a simulation study. Section 1.8 gives an overview of the chapters presented in this book. Section 1.9 provides the conclusion for this chapter.

1.2 TYPES OF SIMULATION

Knowledge about the types of simulation is required to model a system properly. There are two types of simulation models—discrete and continuous. A discrete simulation model has some variables whose values vary at discrete points in time. Some examples of such models are

- Bank tailoring systems
- Railway reservation systems
- Selling systems at counters in supermarkets
- Inventory control systems
- Manufacturing of discrete products

A continuous simulation model has variables that change continuously over time. Thus, the results of such systems are taken at fixed intervals of time after the system reaches a steady state. Examples of such systems include

- Production of chemicals
- Pipeline transmissions of gaseous products
- Medical shops that are open 24 hours

A system can be discrete or continuous depending on the need of the model. For example, a 24-hour medical retail store can be thought of as either a discrete system or a continuous system. If the system is assumed to be discrete, each of the customers' arrivals may be marked as individual arrivals at discrete points in time. On the contrary, if the system is assumed to be continuous in nature, the range of the number of customers can be taken at any time. In addition, most systems are neither perfectly discrete nor perfectly continuous. Most systems are rather a mix of both discrete and continuous systems. Thus, knowledge of both discrete and continuous systems is required.

1.3   STEPS OF SIMULATION

To perform a simulation study properly, the following steps are required to be known, which are also shown graphically in Figure 1.2.

Step 1: Understand the system
   The first step is to understand the system to perform the simulation study. This is an important step since all the other subsequent steps and the success of the simulation study depend on this step. If the system is not known properly, the variables will not be identified properly, and as a result, the measurement of the variables may be erroneous. In addition, understanding a system will reveal the behavior of the system that will help to determine whether a discrete simulation or a continuous simulation is appropriate.
Step 2: Set the goals
   In this step, the goal of the simulation study is set properly. The performance measures are also defined, that is, the parameters and the way to measure performance are defined. The "what-if" scenarios are also defined at this stage. This means that the possible problems are predicted and the actions to be taken to resolve any problems that may occur are decided. For example, the queue discipline (in case of a queue) is decided beforehand.
Step 3: Collect the data
   The relevant input data are collected which may reveal the nature or the inherent distribution of the data. The data may be stationary or nonstationary. Modeling to be done obviously depends on the collected data. For

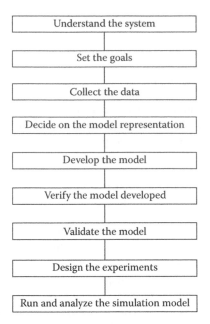

FIGURE 1.2   Steps of simulation study.

example, to determine the distribution of the arrivals of the customers in a system, a significant number of arrivals at different times of a day can be collected. After collecting the data, they may be analyzed for setting a probability distribution that suits the data most. Similarly, the service times also can be measured so as to set a distribution for the service times.

Step 4: Decide on the model representation

In this step, the simulation model to be developed is planned. The design of the model is developed. The development of any model depends on proper designing. If the designing is perfectly done, the development of the model takes reasonable time, and many of the programming hazards can be avoided by this systematic method of designing the system followed by its development. In addition, the software to be used for modeling purpose is chosen. The software may be general-purpose software, such as C, C++, C#, and Java, or proprietary software, such as Arena, AweSim, and ProModel. After choosing the software, the model elements are identified.

Step 5: Develop the model

In this step, the actual simulation model is developed and a test run is performed. Various execution-related parameters, such as the number of runs and the trace elements, are decided.

Step 6: Verify the model developed

In this step, the model developed is verified by checking whether the model is actually representing the system to be simulated and whether the model is producing the desired results. If the observed model does not represent the actual system to be simulated, steps 1–5 are again repeated to check and rectify the components since the developed model did not represent the real system. The results of the developed system should actually match those of the real system in this step.

Step 7: Validate the model

Validation is the most challenging task in simulation. Validation checks whether the results of the system actually match those of the real system. If the results do not match, this would mean that there are errors in any of steps 1–6. In such a case, the process may be repeated from step 3, 4, or 5, depending on the type of error identified, if any.

Step 8: Design the experiments

The questions asked in this step include, for example, "What are the what-if scenarios?" The number of factors or levels is also identified in this step. For example, if there are 3 factors and 2 levels, then a total of $3^2 = 9$ experiments will have to be conducted. Each experiment has to be repeated many times so as to get rid of the effects of uncertainty, to build the confidence level.

Step 9: Run and analyze the simulation model

In this step, the simulation model is run based on the parameters set in steps 1–8. Results are also taken for analysis. Various statistical tools are applied to analyze the results of the simulation study.

1.4 APPLICATION AREAS OF SIMULATION

Nowadays, simulation is applied in almost all fields of science and technology. One of the effective ways to know the recent advancements and the application areas of simulation is to study the research papers published every year in the Winter Simulation Conference (WSC) [2]. The sponsors of the WSC are the American Statistical Association (ASA) (Technical Cosponsor), the Arbeitsgemeinschaft Simulation (ASIM) (Technical Cosponsor), the Association for Computing Machinery: Special Interest Group on Simulation (ACM/SIGSIM), the Institute of Electrical and Electronics Engineers: Systems, Man, and Cybernetics Society (IEEE/SMC) (Technical Cosponsor), the Institute for Operations Research and the Management Sciences: Simulation Society (INFORMS-SIM), Institute of Industrial Engineers (IIE), the National Institute of Standards and Technology (NIST) (Technical Cosponsor), and the Society for Modeling and Simulation International (SCS).

The WSC was started in 1968. The areas of simulation can be obtained from the website of the WSC. A list of areas obtained from the WSC in different years along with some of the topics covered in the WSC is provided in Sections 1.4.1 through 1.4.8.

1.4.1 MANUFACTURING SIMULATION

- Logistics sensitivity of construction processes
- Simulation-based optimization in make-to-order production
- Using a scalable simulation model to evaluate the performance of production system segmentation in a combined material resource planning (MRP) and KANBAN system
- Integration of emulation functionality into an established simulation object library
- Toward an integrated simulation and virtual commissioning environment for controls of material handling systems

- Embedded simulation for automation of material manipulators in a sputtering physical vapor deposition (PVD) process
- Applying semantic web technologies for efficient preparation of simulation studies in production and logistics
- Toward assisted input and output data analysis in manufacturing simulation
- System modeling in SysML and system analysis in Arena
- Simulation and optimization of robot-driven production systems for peak-load reduction
- Real-world simulation-based manufacturing optimization using cuckoo search
- Fast converging, automated experiment runs for material flow simulations using distributed computing and combined metaheuristics
- Model generation in simulation language with extensibility (SLX) using core manufacturing simulation data (CMSD) and extensible markup language (XML) style sheet transformations
- A framework for interoperable sustainable manufacturing process analysis applications development
- Web-based methods for distributing simulation experiments based on the CMSD standard
- Flexible work organization in manufacturing—A simulation-supported feasibility study
- Complex agent interaction in operational simulations for aerospace design

1.4.2 Transport and Logistics Simulation

- Modeling of handling task sequencing to improve crane control strategies in container terminals
- Semiautomatic simulation-based bottleneck detection approach
- Event-based recognition and source identification of transient tailbacks in manufacturing plants
- Statistical modeling of delays in a rail freight transportation network
- Modeling the global freight transportation system
- Simulation backbone for gaming simulation in railways
- Cloud computing architecture for supply chain network simulation
- A simulation-based approach to capturing autocorrelated demand parameter uncertainty in inventory management
- Intra-simulative ecological assessment of logistics networks
- Supply chain carbon footprint trade-offs using simulation
- Approach of methods for increasing flexibility in green supply chains driven by simulation
- Simulation-based optimization heuristic using self-organization for complex assembly lines
- Combining Monte Carlo simulation with heuristics for solving the inventory routing problem with stochastic demands
- Sim-RandSHARP: A hybrid algorithm for solving the arc routing problem with stochastic demands
- Combining point cloud technologies with discrete event simulation

- Automatic collision-free path planning in hybrid triangle and point models
- Assessment methodology for validation of vehicle dynamics simulations using double-lane change maneuver

1.4.3 MILITARY APPLICATIONS

- Effective crowd control through adaptive evolution of agent-based simulation models
- Metamodeling of simulations consisting of time series inputs and outputs
- Assessing the robustness of unmanned aerial vehicle (UAV) assignments
- Agent-based model of the Battle of Isandlwana
- Approximative method of simulating a duel
- Modeling of Canadian Forces' Northern operations and their staging
- North Atlantic Treaty Organization (NATO) Modeling and Simulation Group (MSG)-88 case study results to demonstrate the benefits of using data farming for military decision support
- Joint Coalition Warfighting (JCW) environment development branch support for NATO simulation activities
- Effect of terrain in computational methods for indirect fire
- Effects of stochastic traffic flow model on expected system performance
- International Organization for Standardization (ISO) and Open Geospatial Consortium (OGC) compliant database technology for the development of simulation object databases
- Location model for storage of emergency supplies to respond to technological accidents in Bogota
- Technical combat casualty care: Strategic issues of a serious simulation game development
- Simulating tomorrow's supply chain today

1.4.4 NETWORK SIMULATION

- Runtime performance and virtual network control alternatives in virtual memory (VM)-based high-fidelity network simulations
- Analytical modeling and simulation of the energy consumption of independent tasks
- Validation of application behavior on a virtual time integrated network emulation test bed
- SAFE: Simulation automation framework for experiments
- Simulation visualization of distributed communication systems
- Using network simulation in classroom education

1.4.5 CONSTRUCTION OPERATIONS

- Development of the physics-based assembly system model for the mechatronic validation of automated assembly systems
- Global positioning system (GPS)-based framework toward more realistic and real-time construction equipment operation simulation
- Advancement simulation of tunnel boring machines

- Simulation of crane operation in three-dimensional (3D) space
- Adjusted recombination operator for simulation-based construction schedule optimization
- Intelligent building information modeling (BIM)–based construction scheduling
- Methodology for synchronizing discrete event simulation and system dynamics models
- Construction analysis of rainwater harvesting systems
- Determination of float time for individual construction tasks using constraint-based discrete event simulation
- Transient heat transfer through walls and thermal bridges. Numerical modeling: methodology and validation
- Validation of building energy modeling tools: Ecotect™, Green Building Studio™, and IES<VE>™
- Preliminary research in dynamic-BIM (D-BIM) workbench development

1.4.6 SOCIAL SCIENCE APPLICATIONS

- Equity valuation model of Vietnamese firms in a foreign securities market
- Modeling food supply chains using multiagent simulation
- Hybrid simulation and optimization approach to design and control fresh product networks
- Hypercube simulation analysis for a large-scale ambulance service system
- Studying the effect of mosque configuration on egress times
- An open source simulation-based approach for neighborhood spatial planning policy
- Modeling social groups in crowds using common ground theory
- Grounded theory-based agent
- Agent-based model of the E-MINI future market: Applied to policy decisions
- Predictive nonequilibrium social science
- Do the attributes of products matter for success in social network markets?
- Complexity and agent-based models in the policy process
- Modeling innovation networks of general purpose technologies—The case of nanotechnology
- A generic model to assess sustainability impact of resource management plans in multiple regulatory contexts
- Using participatory elicitation to identify population needs and power structures in conflict environments
- Peer review under the microscope: An agent-based model of scientific collaboration

1.4.7 ENVIRONMENT APPLICATIONS

- Simulation to discover structure in optimal dynamic control policies
- Material flow cost accounting (MFCA)–based simulation analysis for environment-oriented supply chain management (SCM) optimization conducted by small and medium sized enterprises (SMEs)

- Using discrete event simulation to evaluate a new master plan for a sanitary infrastructure
- Experiences with object-oriented and equation-based modeling of a floating support structure for wind turbines in Modelica
- A comparative analysis of decentralized power grid stabilization strategies
- A hybrid simulation framework to assess the impact of renewable generators on a distribution network
- Achieving sustainability through a combination of life cycle analysis (LCA) and discrete event simulation (DES) integrated in a simulation software for production processes
- Evaluation of methods used for life-cycle assessments in discrete event simulation
- Global sensitivity analysis of nonlinear mathematical models—An implementation of two complementing variance-based methods
- Large-scale traffic simulation for low-carbon city
- Simulation-based validity analysis of ecological user equilibrium
- Cellular automata model based on machine learning methods for simulating land use change

1.4.8 HEALTH CARE APPLICATIONS

- A survey on the use of simulation in German health care
- Why health care professionals are slow to adopt modeling and simulation
- SimPHO: An ontology for simulation modeling of population health
- Applying a framework for health care incentives simulation
- A generalized simulation model of an integrated emergency post
- Mixing other methods with simulation is no big deal
- Hybrid simulation for modeling large systems
- Modeling the spread of community-associated methicillin-resistant *Staphylococcus aureus* (MRSA)
- High performance informatics for pandemic preparedness
- A large simulation experiment to test influenza pandemic behavior
- Hybrid simulation with loosely coupled system dynamics and agent-based models for prospective health technology assessments
- Calibration of a decision-making process in a simulation model by a bicriteria optimization problem
- Modeling requirements for an emergency medical service system design evaluator
- A simulation-based decision support system to model complex demand-driven health care facilities
- A simulation study to reduce nurse overtime and improve patient flow time at a hospital endoscopy unit
- A simulation study of patient flow for the day of surgery admission
- A multi-paradigm, whole system view of health and social care for age-related macular degeneration
- Linked lives: The utility of an agent-based approach to modeling partnership and household formation in the context of social care

- Using system dynamics to model the social care system: Linking demography, simulation, and care
- Evaluating health care systems with insufficient capacity to meet the demand
- The case against utilization: Deceptive performance measures in in-patient care capacity models
- Planning of bed capacities in specialized and integrated care units incorporating bed blockers in a simulation of surgical throughput
- A simulation-based iterative method for a trauma center—Air ambulance location problem
- Reducing ambulance response time using simulation: The case of Val-de-Marne department emergency medical service
- Comparison of ambulance diversion policies via simulation
- Operations analysis and appointment scheduling for an outpatient chemotherapy department
- Sensitivity analysis of an intensive care unit (ICU) simulation model
- Aggregate simulation modeling of an magnetic resonance imaging (MRI) department using effective process times
- ABMS optimization for emergency departments
- Multi-criteria framework for emergency department in Irish hospital
- Simulation with data scarcity: Developing a simulation model of a hospital emergency department

1.5   SIMULATION OF QUEUING SYSTEMS

In this section, an example of simulation for a queuing system is given. In particular, the Monte Carlo simulation method has been applied. The interarrival times of customers at a queue and service times are given in Table 1.1.

We can simulate the queuing behavior based on the data in Table 1.1. Thus, the cumulative probabilities are first calculated from the given data, and accordingly, the intervals for random numbers are decided as shown in Table 1.2.

Next, a set of random numbers can be generated. The number of generated random numbers may be taken to be equal to that of customer arrivals to simulate. Suppose a total of 20 arrivals are simulated. Table 1.3 shows the generated random numbers and the related calculations. Let the system start at 9:00 a.m.

TABLE 1.1
Interarrival and Service Times

Interarrival Times		Service Times	
Minutes	Probability	Minutes	Probability
2	0.10	1	0.15
4	0.25	3	0.17
6	0.35	5	0.30
8	0.20	7	0.28
10	0.10	9	0.10

TABLE 1.2
Random Number Coding

Arrival Times				Service Times			
Minutes	Probability	Cumulative Probability	Random Number Interval	Minutes	Probability	Cumulative Probability	Random Number Interval
2	0.10	0.10	00–09	1	0.15	0.15	00–14
4	0.25	0.35	10–34	3	0.17	0.32	15–31
6	0.35	0.70	35–69	5	0.30	0.62	32–61
8	0.20	0.90	70–89	7	0.28	0.90	62–89
10	0.10	1.00	90–99	9	0.10	1.00	90–99

TABLE 1.3
Relevant Calculation for Queuing System

Random Number	Interarrival Time (minutes)	Arrival Time (a.m.)	Service Starts At (a.m.)	Random Number	Service Time (minutes)	Service Ends At (a.m.)	Waiting Time of Customers in Queue (minutes)	Idle Time of Server (minutes)
94	10	09:10	09:10	36	5	09:15	0	10
73	8	09:18	09:18	46	5	09:23	0	3
78	8	09:26	09:26	34	5	09:31	0	3
72	8	09:34	09:34	70	7	09:41	0	3
59	6	09:40	09:41	97	9	09:50	1	0
63	8	09:48	09:50	80	7	09:57	2	0
85	6	09:54	09:57	40	5	10:02	3	0
66	6	10:00	10:02	94	9	10:11	2	0
61	6	10:06	10:11	55	5	10:16	5	0
23	4	10:10	10:16	43	5	10:21	6	0
71	8	10:18	10:21	15	3	10:24	3	0
26	4	10:22	10:24	67	7	10:31	2	0
96	10	10:32	10:32	78	7	10:39	0	1
73	8	10:40	10:40	21	3	10:43	0	1
77	8	10:48	10:48	22	3	10:51	0	5
9	2	10:50	10:51	41	5	10:56	1	0
14	4	10:54	10:56	35	5	11:01	2	0
88	8	11:02	11:02	87	7	11:09	0	1
64	6	11:08	11:09	35	5	11:14	1	0
82	8	11:16	11:16	29	3	11:19	0	2
Total							28	29

Table 1.3 shows that the total waiting time of the customers is 28 minutes and the total idle time of the service clerk is 29 minutes. Thus, the average waiting time of the customers is 1.40 minutes and the average idle time of the service clerk is 1.45 minutes.

1.6   SIMULATION OF INVENTORY SYSTEM

Tables 1.4 and 1.5 show the demands and their respective probabilities, and the lead times and their respective probabilities, respectively.

Based on these tables, the inventory system can be simulated. Suppose the maximum inventory level is 14 units and the review period is 6 days. Table 1.6 gives the inventory status for five cycles. Assume that a previous order of 10 units will arrive on day 2 of cycle 1.

Table 1.6 shows that the order of 10 units arrives on day 2, and as a result, the new inventory level is $10 + 2 = 12$ units, from which the demand of 1 unit for day 2 is satisfied, resulting in the ending inventory of 11 units, which is also the beginning inventory of the next day. From this inventory, the demands of the remaining 5 days of the cycle are satisfied. On day 6, there is a shortage of 1 unit since the demand is greater by 1 unit than the inventory on hand. Since day 6 is the last day of the cycle, the order is placed on this day. Since the maximum inventory allowed is 14 units and there is a shortage of 1 unit, a total order of $14 + 1 = 15$ units is placed. The random number for the lead time generates a lead time of 0 day, which means that the inventory is replenished on day 1 of cycle 2. After satisfying the shortage of 1 unit, the inventory level turns out to be 14 units on day 1 of cycle 2. Note that this 14 units of inventory is the maximum inventory allowed.

TABLE 1.4
Probabilistic Daily Demands

Demand	Probability	Cumulative Probability	Random Number Intervals
0	0.10	0.10	00–09
1	0.20	0.30	10–29
2	0.25	0.55	30–54
3	0.22	0.77	55–76
4	0.12	0.89	77–88
5	0.11	1.00	89–99

TABLE 1.5
Probabilistic Lead Time

Lead Time	Probability	Cumulative Probability	Random Number Intervals
0	0.2	0.20	00–19
1	0.3	0.50	20–49
2	0.4	0.90	50–89
3	0.1	1.00	90–99

TABLE 1.6

Calculation of Inventory Levels Based on Random Demand

Cycle	Day	Beginning Inventory	Random Number	Demand	Ending Inventory	Inventory Shortage	Order Quantity	Random Number	Days to Arrive Orders
1	1	6	82	4	2	0	0	–	0
	2	12	25	1	11	0	0	–	–
	3	11	85	4	7	0	0	–	–
	4	7	98	5	2	0	0	–	–
	5	2	21	1	1	0	0	–	–
	6	1	40	2	0	1	15	9	0
2	1	14	16	1	13	0	0	–	–
	2	13	54	2	11	0	0	–	–
	3	11	71	3	8	0	0	–	–
	4	8	45	2	6	0	0	–	–
	5	6	86	4	2	0	0	–	–
	6	2	34	2	0	0	14	92	3
3	1	0	62	3	0	3	0	–	2
	2	0	25	1	0	4	0	–	1
	3	0	30	2	0	6	0	–	0
	4	8	73	3	5	0	0	–	–
	5	5	31	2	3	0	0	–	–
	6	3	78	4	0	1	15	21	1
4	1	0	45	2	0	3	0	–	0
	2	12	97	5	7	0	0	–	–
	3	7	21	1	6	0	0	–	–
	4	6	38	2	4	0	0	–	–
	5	4	63	3	1	0	0	–	–
	6	1	53	2	0	1	15	57	2
5	1	0	59	3	0	4	0	–	1
	2	0	72	3	0	7	0	–	0
	3	8	47	2	6	0	0	–	–
	4	6	59	3	3	0	0	–	–
	5	3	10	1	2	0	0	–	–
	6	2	70	3	0	1	15	67	2

Cycle 2 runs well without any shortage, but shortage of inventory is observed on day 1 of cycle 3, and this shortage is accumulated as the days of the cycle pass; the total shortage of 6 units is satisfied after the arrival of the items at inventory on day 4 of cycle 3, since the lead time for the order placed at the end of the previous cycle is 3 as generated from the random number for the lead time. In this way, the calculations for a total of five cycles are done. The average inventory for the five cycles is $100/30 = 3.33$, since the total ending inventory is 100 units for 30 days. The average shortage of inventory for 30 days is $31/30 = 1.033$, since the total shortage is 31 units for 30 days.

1.7   ADVANTAGES AND DISADVANTAGES OF SIMULATION

Simulation is a modeling by which a large system can be analyzed, thereby saving a lot of money. This is the biggest advantage of simulation. However, the other advantages of simulation are listed as follows:

1. Large systems can be analyzed with ease with the help of simulation.
2. Simulation saves money by canceling the need to observe and analyze the larger system.
3. Simulation study uses real data, and therefore the results obtained from simulation studies are also very close to reality.
4. Conditions of a simulation experiment can easily be varied and the resulting outcomes can also be investigated very easily.
5. Critical situation can be investigated through simulation without risk.
6. Simulation experiment can be lingered so that the behavior of a system can be studied over longer period.
7. The speed of simulation can be changed to study the system closely.

The disadvantages of simulation are listed as follows:

1. The initial cost of installing a simulation system may be high.
2. It can be costly to measure how one variable affects the other.
3. To conduct an effective simulation study, the concerned person needs to have a thorough understanding of the actual physical system, and a knowledge about the various parameters of the system is also required.

In spite of having some difficulties, simulation is always preferred over the direct observation of the actual physical system since simulation is risk free.

1.8   OVERVIEW OF THE REMAINING CHAPTERS

This book describes the various aspects of simulation. Thus, all the chapters have been arranged accordingly. An overview of each of the chapters is briefly discussed in this section.

This chapter is an introductory chapter for simulation. Thus, this chapter mainly provides the definition and discusses other related issues concerned with simulation. A total of two examples of simulations have been provided in this chapter. The first example is the queuing system and the second example is the inventory system. The definition of simulation is provided in Section 1.1. The types of simulation are also described in this chapter. Lists of advantages and disadvantages are provided in Section 1.7.

Chapter 2 discusses Monte Carlo simulation, which is the basic and traditional form of simulation. A brief introduction to Monte Carlo techniques and variance reduction techniques has been provided. In addition, the application areas of Monte Carlo simulation are also provided.

Chapter 3 provides an introduction to general probability theory. A brief description of various ways to calculate the probability values is presented. Among the various methods, Bayes' theorem is a vital one. A number of numerical examples are provided to explain the calculation of the probability values using the given expressions.

Chapter 4 discusses various discrete and continuous probability distributions. The discrete probability distributions described are discrete uniform distribution, binomial distribution, geometric distribution, negative binomial distribution, hypergeometric distribution, and Poisson distribution. The continuous probability distributions discussed are exponential distribution, beta distribution, gamma distribution, normal distribution, lognormal distribution, triangular distribution, Weibull distribution, and Erlang distribution.

Chapter 5 describes various random number generators that are very essential for the simulation study. The various pseudorandom number generators are linear congruential generator, multiplicative congruential generator, inverse congruential generator, combined linear congruential generator, lagged Fibonacci generator, and midsquare generator. The special pseudorandom number generators discussed are Mersenne Twister algorithm and Marsaglia generator. In addition, the quasi-random numbers are also briefly discussed.

Chapter 6 discusses the generation of four random variates. The four methods discussed are inverse transform technique, convolution method, acceptance–rejection method, and method of composition. The first three methods are discussed in detail and given special emphasis.

Chapter 7 discusses the steady-state behavior of the stochastic processes. Various stochastic processes are discussed in this chapter. The particular stochastic processes discussed are Bernoulli process, simple random walk, population process, stationary process, autoregressive process, Markov process, and Poisson process.

Chapter 8 briefly discusses the steady-state parameters of stochastic processes. The terminating simulation and the steady-state simulation and their scopes are particularly discussed.

Chapter 9 represents the various aspects of computer simulation. Various views on computer simulations discussed in the existing literature along with some other types of simulations are also discussed. Particular distinction is made between simulation of computer systems and computer simulation of various fields of study.

Chapter 10 concentrates on manufacturing simulation, that is, simulation in the field of manufacturing. Various aspects of manufacturing simulation are provided. The direction of how to select particular manufacturing simulation software and a list of manufacturing simulation software is presented in this chapter. A brief introduction to Arena simulation software is also provided.

Chapter 11 mainly describes some of the general-purpose and proprietary simulation software. The general-purpose simulation software discussed are C and C++. The proprietary software discussed are AweSim for manufacturing system and beer distribution game simulation software for supply chain. This chapter is particularly dedicated to manufacturing and supply chain since these two areas are the main emphasis of this book, and various research papers on simulation are published in these two areas.

Chapter 12 emphasizes on simulation in supply chain. Various aspects of supply chain simulation are presented in this chapter. Various types of simulation technologies, for example, beer distribution game simulation and simulation with system dynamics, are also presented.

Chapter 13 gives a brief glimpse of the simulation in various other fields of study. Particular emphasis is given to electrical engineering, chemical engineering, aerospace engineering, and civil engineering.

Chapters 14 and 15 are devoted to complex simulation, that is, simulation of complex systems. One of the ways to simulate complex systems is the use of automata. The simulation of cellular automata is briefly discussed in Chapter 15.

Chapter 16 presents a detailed study on agent-based simulation. The characteristics of agents, the types of agents, the phases of general agent-based simulation, the design methodology of agent-based systems, and multiagent-based manufacturing systems are presented in this chapter.

Chapter 17 is devoted to continuous simulation. In particular, various calculus-based methods are applied in case of continuous simulation. Thus, the integration-based methods and various validation schemes are also described.

Chapters 18 and 19 discuss simulation optimization, which is a very advanced topic in the study of simulation. Various aspects of continuous simulation are presented in Chapter 18, and various simulation optimization algorithms proposed in the existing literature are presented in Chapter 19.

Chapter 20 is a special chapter written on simulation with system dynamics. System dynamics is especially applied for behavioral simulation of the supply chain and other related areas of study.

Chapter 21 describes the various aspects of simulation software, which include the various methods of selecting and evaluating the simulation software evident from the existing literature.

Chapter 22 is the last chapter that discusses the future technologies that are expected to be used in future simulation technologies. Particular emphasis is given to .NET technologies and cloud computing.

1.9   CONCLUSION

This chapter is a brief introduction to this book on simulation. This chapter provided some overview of the simulation definition and the related aspects of simulation. The concepts of system and model have been explained. Then, the type of experiments and the type of simulation have been depicted. An example of simulation of a queuing system has been cited. The advantages and disadvantages of simulation have also been presented.

REFERENCES

1. Law, A. M. and Kelton, W. D. (2003). *Simulation Modeling and Analysis*. 3rd Edition. New Delhi, India: Tata McGraw-Hill.
2. http://informs-sim.org/or http://www.wintersim.org.

2 Monte Carlo Simulation

2.1 INTRODUCTION

A simulation model can be deterministic or stochastic in nature. In a deterministic simulation model, the variables (both input and output variables) are not random variables, whereas in a stochastic simulation model, one or more variables are stochastic in nature. Monte Carlo simulation is basically a stochastic simulation method and a sampling technique in which random sampling is basically used.

An early version of Monte Carlo simulation was observed in Buffon's needle experiment in which the value of π could be calculated by dropping a needle on a floor that was made by several strips of wood. The real Monte Carlo simulation was introduced first by Enrico Fermi [1] in 1930 while studying neutron diffusion. The word "Monte Carlo" was first coined by Von Neumann during the 1940s, that is, during World War II.

Monte Carlo simulation is used to solve problems that are difficult to represent by an exact mathematical model or problems in which the solution is difficult to obtain by a simple analytical method. Monte Carlo simulation uses a user-defined probability distribution for any uncertain variable in a problem. For this reason, a Monte Carlo model is said to be a static method rather than a dynamic method. Thus, Monte Carlo simulation generates random numbers and uses these numbers for a variety of problems, such as forecasting, risk analysis, and other estimations. Another important application of the Monte Carlo technique is to evaluate difficult integrals. Section 2.4 shows such evaluation through the use of various Monte Carlo methods. Consider, for example, the two examples of Monte Carlo simulation problems in Section 2.1.1.

2.1.1 EXAMPLES

In this section, we show two examples to explain the use of Monte Carlo simulation in various real-world problems.

Example 1

ABC bakery keeps stock of a popular cake. The daily demands gathered from the experience of the manager are given in Table 2.1.

The following random numbers have been generated to get an approximate estimate of demand through simulation of demand for 15 days:

38   77   42   58   62   81   25   56   92   8   30   47   9   54   63

TABLE 2.1

Demands and Respective Probabilities for Example 1

Demands	0	10	20	30	40	50
Probabilities	0.02	0.30	0.05	0.40	0.15	0.08

In this problem, we have user-defined random numbers on the basis of which we will have to forecast the demand. To forecast the demand, we need to use the probability values given in Table 2.2.

The range of random numbers will be used for forecasting the demand (Table 2.3) based on the given set of predefined random numbers obtained from the users.

To get the forecast demand, every random number is matched with the range of random numbers in Table 2.2 and the respective demand is found to be the forecast demand. For example, in the first row of Table 2.3, the generated random number is 38, which is matched with the range of random numbers listed in the fourth column of Table 2.2. The identified range comes out to be 37–76 since the number 38 is contained in this range. The respective demand for the

TABLE 2.2

Range of Random Numbers for Example 1

Demands	Probabilities	Cumulative Probabilities	Range of Random Numbers
0	0.02	0.02	00–01
10	0.30	0.32	02–31
20	0.05	0.37	32–36
30	0.40	0.77	37–76
40	0.15	0.92	77–91
50	0.08	1.00	92–99

TABLE 2.3

Forecasting of Demands for Example 1

Given Random Numbers	Forecast Demands	Given Random Numbers	Forecast Demands
38	30	92	50
77	40	8	10
42	30	30	10
58	30	47	30
62	30	9	10
81	40	54	30
25	10	63	30
56	30		

range 37–76 is 30, which is the forecast demand against the generated random number 38, as shown in Table 2.3. The generated forecast demands for 15 days can be used for various purposes such as to determine the stock level for the next 15 days.

Example 2

Consider a queuing system in a bank. Suppose that the bank under consideration needs to analyze the arrival pattern of its customers and the service pattern of its internal clerk during a certain time period of a day. The service manager has generated the probabilities for interarrival and service times that are given in Table 2.4.

Since both the arrival and the service are random in nature, a set of 40 numbers has been generated to make an estimate of the random nature of the system. The generated 40 random numbers are given as follows:

79 24 32 20 82 14 3 44 57 61 35 51 93 73 9 16 22 45 64 76 87

19 60 20 32 66 56 41 97 71 86 54 17 64 53 5 81 67 38 80

Thus, the simulation will have to be carried out on the basis of these 40 random numbers. Tables 2.5 and 2.6 calculate the ranges of random numbers.

Thus, applying the random numbers as earlier, we get the uncertain interarrival times and the uncertain service times. Table 2.7 provides estimates of the service times and interarrival times obtained from the given random numbers.

TABLE 2.4

Interarrival and Service Times for Example 2

Interarrival Times	Probabilities	Service Times	Probabilities
1	0.24	2	0.12
3	0.18	3	0.34
5	0.29	4	0.22
7	0.19	5	0.17
9	0.10	6	0.15

TABLE 2.5

Range of Random Numbers for Interarrival Times

Interarrival Times	Probabilities	Cumulative Probabilities	Range of Random Numbers
1	0.24	0.24	00–23
3	0.18	0.42	24–41
5	0.29	0.71	42–70
7	0.19	0.90	71–89
9	0.10	1.00	90–99

TABLE 2.6
Range of Random Numbers for Service Times

Service Times	Probabilities	Cumulative Probabilities	Range of Random Numbers
2	0.12	0.12	00–11
3	0.34	0.46	12–45
4	0.22	0.68	46–67
5	0.17	0.85	68–84
6	0.15	1.00	85–99

TABLE 2.7
Estimate of Interarrival Times and Service Times

Random Numbers	Interarrival Times	Service Times	Random Numbers	Interarrival Times	Service Times
79	7	5	87	7	6
24	3	3	19	1	3
32	3	3	60	5	4
20	1	3	20	1	3
82	7	5	32	3	3
14	1	3	66	5	4
3	1	2	56	5	4
44	5	3	41	3	3
57	5	4	97	9	6
61	5	4	71	7	5
35	3	3	86	7	6
51	5	4	54	5	4
93	9	6	17	1	2
73	7	5	64	5	4
9	1	1	53	5	4
16	1	3	5	1	1
22	1	3	81	7	5
45	5	3	67	5	4
64	5	4	38	3	3
76	7	5	80	7	5

Note that the same set of random numbers has been used for both the interarrival times and the service times.

Suppose that the first customer enters the bank for service at 10:00 a.m. Thus, the behavior of the queuing system is depicted in Table 2.8.

From the above problems, the method of applying the Monte Carlo simulation can easily be identified. The Monte Carlo simulation procedure can be delineated

TABLE 2.8
Behavior of Queue in Bank

Customer	Interarrival Time (minutes)	Customer Enters At	Service Time	Service Starts At	Service Ends At	Waiting Time of Clerk	Waiting Time of Customer
1	7	10:07 a.m.	5	10:07 a.m.	10:12 a.m.	7	–
2	3	10:10 a.m.	3	10:12 a.m.	10:15 a.m.	–	2
3	3	10:13 a.m.	3	10:15 a.m.	10:18 a.m.	–	2
4	1	10:14 a.m.	3	10:18 a.m.	10:21 a.m.	–	4
5	7	10:21 a.m.	5	10:21 a.m.	10:26 a.m.	–	–
6	1	10:22 a.m.	3	10:26 a.m.	10:29 a.m.	–	4
7	1	10:23 a.m.	2	10:29 a.m.	10:31 a.m.	–	6
8	5	10:28 a.m.	3	10:31 a.m.	10:34 a.m.	–	3
9	5	10:33 a.m.	4	10:34 a.m.	10:38 a.m.	–	1
10	5	10:38 a.m.	4	10:38 a.m.	10:42 a.m.	–	–
11	3	10:41 a.m.	3	10:42 a.m.	10:45 a.m.	–	1
12	5	10:46 a.m.	4	10:46 a.m.	10:50 a.m.	1	–
13	9	10:55 a.m.	6	10:55 a.m.	11:01 a.m.	5	–
14	7	11:02 a.m.	5	11:02 a.m.	11:07 a.m.	1	–
15	1	11:03 a.m.	1	11:07 a.m.	11:08 a.m.	–	1
16	1	11:04 a.m.	3	11:08 a.m.	11:11 a.m.	–	4
17	1	11:05 a.m.	3	11:11 a.m.	11:14 a.m.	–	6
18	5	11:10 a.m.	3	11:14 a.m.	11:17 a.m.	–	4
19	5	11:15 a.m.	4	11:17 a.m.	11:21 a.m.	–	2
20	7	11:22 a.m.	5	11:22 a.m.	11:27 a.m.	1	–
21	7	11:29 a.m.	6	11:29 a.m.	11:35 a.m.	2	–
22	1	11:30 a.m.	3	11:35 a.m.	11:38 a.m.	–	5
23	5	11:35 a.m.	4	11:38 a.m.	11:42 a.m.	–	3
24	1	11:36 a.m.	3	11:42 a.m.	11:45 a.m.	–	6
25	3	11:39 a.m.	3	11:45 a.m.	11:48 a.m.	–	6
26	5	11:44 a.m.	4	11:48 a.m.	11:52 a.m.	–	4
27	5	11:49 a.m.	4	11:52 a.m.	11:56 a.m.	–	3
28	3	11:52 a.m.	3	11:56 a.m.	11:59 a.m.	–	4
29	9	12:01 p.m.	6	12:01 p.m.	12:07 p.m.	2	–
30	7	12:08 p.m.	5	12:08 p.m.	12:13 p.m.	1	–
31	7	12:15 p.m.	6	12:15 p.m.	12:21 p.m.	2	–
32	5	12:20 p.m.	4	12:21 p.m.	12:25 p.m.	–	1
33	1	12:21 p.m.	2	12:25 p.m.	12:27 p.m.	–	4
34	5	12:26 p.m.	4	12:27 p.m.	12:31 p.m.	–	1
35	5	12:31 p.m.	4	12:31 p.m.	12:35 p.m.	–	–
36	1	12:32 p.m.	1	12:33 p.m.	12:36 p.m.	–	3
37	7	12:39 p.m.	5	12:39 p.m.	12:44 p.m.	3	–
38	5	12:44 p.m.	4	12:44 p.m.	12:48 p.m.	–	–

(Continued)

TABLE 2.8

(Continued) Behavior of Queue in Bank

Customer	Interarrival Time (minutes)	Customer Enters At	Service Time	Service Starts At	Service Ends At	Waiting Time of Clerk	Waiting Time of Customer
39	3	12:47 p.m.	3	12:48 p.m.	12:51 p.m.	–	1
40	7	12:54 p.m.	5	12:54 p.m.	12:59 p.m.	3	–

The average waiting time of the clerk is found out to be the total waiting time of the clerk/the total number of customers (simulation runs), that is, 28/40 = 0.7 minute. Similarly, the average service time is the total service time/the total number of customers (simulation runs), that is, 81/40 = 2.05 minutes.

in Section 2.1. The remaining sections of this chapter are arranged as follows: Section 2.2 shows the steps of Monte Carlo simulation. Section 2.3 introduces the methods of generating random numbers. Section 2.4 describes the types of Monte Carlo simulation. Section 2.5 depicts the various variance reduction techniques. Section 2.6 demonstrates when to use Monte Carlo simulation. Section 2.7 mentions some application areas of Monte Carlo simulation. Section 2.8 lists the advantages and disadvantages of Monte Carlo simulation. Section 2.9 provides the conclusion for this chapter.

2.2   STEPS OF MONTE CARLO SIMULATION

The basic steps of the Monte Carlo simulation are provided as follows:

Step 1: Define the problem and identify the factors that impact the objectives of the problem.
Step 2: Identify the variables of the problem on which the simulation will have to be performed.
Step 3: Decide on the rules on which the simulation will have to be conducted.
Step 4: Decide on the starting condition of the simulation experiment.
Step 5: Decide on the number of simulation runs to be performed.
Step 6: Select a random number generator and generate the random numbers by the generator.
Step 7: Associate the random number with the factor identified.
Step 8: Examine and evaluate the results of the experimentation.

The above steps can be explained in light of the given examples. Let us consider the queuing example of Example 2. The objective of the problem is to reduce the waiting time of the customers and the waiting time of the system as a whole. The waiting time of the system is the summation of the waiting time of the customer and the waiting time of the clerk. From the objectives, the variables are clearly identified as "interarrival times" and "service times." The rule of the service, as assumed here, is first-come, first-served-based service, which defines the queue discipline. Thus, a customer arrives at the system, waits in the queue for the service, and then receives the service by the clerk. The bank is assumed to open at 10:00 a.m., and the first customer arrives at 10:07 a.m. following the first interarrival time. The probability values for the possible interarrival and service times are also generated based on the past experience by the manager of the bank. A total of 40 arrivals are assumed. The random numbers for the interarrival times and service times have already been decided earlier. With the help of the generated random numbers, the simulation of the interarrival and service times is performed. These values are then utilized to calculate the waiting times of both the customers and the clerk.

The above steps and problems clearly show that the success and the effectiveness of Monte Carlo simulation largely depend on the generated random numbers. In addition, Monte Carlo simulation is basically a computerized or automated simulation procedure. Thus, the method of generating random numbers influences the quality of the Monte Carlo simulation. Computer-generated random numbers fall into two categories: pseudorandom numbers (PRNs) and chaotic random numbers [2]. Thus, Section 2.3 provides a brief introduction to these two types of random numbers. However, the random number generation will be discussed in detail in Chapter 5.

2.3   RANDOM NUMBER GENERATORS

Random numbers can be generated by the computer depending on the use of the PRNs or the chaotic random numbers. After generating the random numbers, various variance reduction techniques can be used to improve the chosen random number generator.

The common PRN generators are a linear congruential method and a lagged Fibonacci method, among many others. The respective expressions for these two generators are given in the following expressions:

$$x_n = cx_{n-1} \bmod M \tag{2.1}$$

$$x_n = x_{n-p} \odot x_{n-q} \bmod M \tag{2.2}$$

where:
   M is a large integer (say 2^{31})
   c is a constant
   p and q are large prime numbers
   $\odot$ is a binary operator such as addition, subtraction, multiplication, or XOR operator

The seed value for both these methods can be taken from the random number table. The generator random numbers must be tested for randomness after their generation. The best tests, as evident from the existing literature, are found to be the lattice test and the spectral test. Some of the other tests include phase plots, spatial plots, histograms, periodograms, and so on.

A chaotic system is basically a dynamic system that depends on the initial condition. If these initial conditions and parameters are chosen wisely, it can be very ergodic and unpredictable in nature, and by doing so, it becomes a very good random number generator. Appropriate transformation functions may be used to make the generated random numbers exhibit a uniform distribution property.

One of the best-known chaotic maps is generated by the logistic expression

$$f(x) = 4x(1 - x) \tag{2.3}$$

which is quadratic in nature. The parameter 4 can be changed to show the different behaviors of the random numbers. If applied repeatedly for any starting value (0,1), it gives a map in (0,1), which implies the ergodicity of the process and hence its apparent randomness.

For a uniformly distributed chaotic random number sequence, the following transformation may be used:

$$\int_0^1 f(u)du = \int_0^1 f(x)\left|\frac{du}{dx}\right|dx = \int_0^1 f(x)\pi\sqrt{x(1-x)}dx \tag{2.4}$$

Correspondingly, in the discrete Monte Carlo simulation over (0,1), we have

$$\frac{1}{N}\sum_{n=1}^N f(u_n) = \frac{\pi}{N}\sum_{n=1}^N f(x_n)\sqrt{x_n(1-x_n)} \tag{2.5}$$

Thus, the recursive expression based on expression (2.3) can be given by the following:

$$x_{n+1} = 4x_n(1-x_n) \tag{2.6}$$

2.4   TYPES OF MONTE CARLO SIMULATIONS

This section discusses four different Monte Carlo techniques that can also be used to solve various mathematical expressions. The numbers generated by evaluating the mathematical expression can be used as random numbers. The four Monte Carlo techniques are as follows:

1. Crude Monte Carlo
2. Acceptance–rejection Monte Carlo
3. Stratified sampling
4. Importance sampling

As mentioned in Section 2.1, all of the above four techniques will be discussed in light of integration of difficult integrals with these techniques. The integral considered to solve is provided as follows:

$$I = \int_a^b f(x)dx \tag{2.7}$$

2.4.1 CRUDE MONTE CARLO

The basic concept of this method is to take N random samples first. For each random sample s, the function $f(s)$ is calculated. Then, all the values for all the samples are summed up, and the result is divided by N to get the average value. Then, this average value is multiplied by $(b-a)$ to get the evaluated integral. This entire operation can be expressed by the following equation:

$$\bar{\lambda} = \frac{(b-a)}{N} \sum_{i=1}^{N} Z_i \tag{2.8}$$

where:
   Z_i is an estimator of the function $f(X_i)$

To verify the accuracy of the estimate, the variance of the samples can be calculated, which can be used to determine the confidence interval of the method. The calculation steps for this method are given as follows:

Step 1: Consider the evaluation of the integral $I = \int_a^b f(x)\mathrm{d}x$.

Step 2: Generate the uniformly distributed random variables $X_1, X_2, \ldots,$ $X_N \sim U(a,b)$ from the uniformly distributed random variables $Y_1, Y_2, \ldots,$ $Y_N \sim U(0,1)$.

Step 3: Suppose Z_i is an indicator function for a pair of random numbers (X_i, Y_i) that lies below the curve represented by $f(x)$. Then the value of Z_i is defined as follows:

$$Z_i = \begin{cases} 1, & \text{if } Y_i < f(X_i) \\ 0, & \text{otherwise} \end{cases} \tag{2.9}$$

Step 4: Define $\lambda = E(Z_i)(b-a)$, where λ is an estimator of the integral I.

Step 5: Thus, the mean estimate $\bar{\lambda}$ is given by the following:

$$\bar{\lambda} = \frac{(b-a)}{N} \sum_{i=1}^{N} Z_i$$

The following numerical example will make the steps clearer.

Example 3

Let us consider the integral $I = \int_a^b f(x)\mathrm{d}x = \int_1^2 (1/x)\mathrm{d}x = \ln(2)$. The crude Monte Carlo method can be used to approximate the value of $\ln(2)$. Table 2.9 shows the values of $\ln(2)$ with different values of N.

TABLE 2.9
Value of ln(2) with Different Values of N

N	ln(2)
1	1.000
2	1.000
4	0.500
8	0.875
16	0.562
32	0.625
64	0.719
128	0.633
256	0.648
512	0.691
1,024	0.700
2,048	0.677
4,096	0.689
8,192	0.691
16,384	0.688
32,768	0.691
65,536	0.692
131,072	0.694
262,144	0.692
524,288	0.692
1,048,576	0.693

2.4.2 ACCEPTANCE–REJECTION MONTE CARLO

The second type of Monte Carlo technique is the acceptance–rejection Monte Carlo method. The basic method for this technique is easy, but it is also the least accurate.

The basic idea is simple. For each interval (a,b) and for a given value of x, we can find the upper limit. We then enclose this interval with a rectangle whose height is above the upper limit, so that the entire function for this particular interval (a,b) falls within the rectangle. Next, we can take random points from within the rectangle and check to see whether the point is below the curve represented by $f(x)$. A point is a successful one if it is below the curve. Once this sampling is finished, we can approximate $f(x)$ for (a,b) by finding out the area of the surrounding rectangle. Now, this area can be multiplied by the number of successful samples, and the average is taken by dividing by the total N number of samples. This will evaluate the approximate value of the integral. Thus, the ratio of the entire area below $f(x)$ and the entire area of the rectangle is actually the ratio of the successful samples k and N.

Thus, mathematically, we have the following:

$$\int_a^b f(x)\mathrm{d}x \approx \frac{k}{N}M(b-a) \qquad (2.10)$$

The accuracy of this method can again be verified by calculating the variance of the samples.

2.4.3 STRATIFIED SAMPLING

Stratified sampling is the third Monte Carlo technique. The basic idea is to divide the interval (a,b) into a number of subintervals. Then, the crude Monte Carlo is applied to each of the subintervals, and the results are added to get the approximate value of the integral. Mathematically, this can be expressed as follows:

$$\int_a^b f(x)\mathrm{d}x = \int_a^c f(x)\mathrm{d}x + \int_c^b f(x)\mathrm{d}x \tag{2.11}$$

Here, the interval (a,b) has been broken into two intervals, (a,c) and (c,b), and crude Monte Carlo has been applied on each of these subintervals.

2.4.4 IMPORTANCE SAMPLING

Importance sampling is the most complex technique among all the Monte Carlo techniques. It is based on stratified sampling. Like stratified sampling, importance sampling also subdivides the interval (a,b). But after subdividing, the results of each of the subdivisions are observed. Next, the subdivisions with higher integral values, that is, the important subdivisions, are identified. These more important subdivisions are then further subdivided and ultimately all the results are added to get the approximate integral. By emphasizing the important subintervals, more accurate results may be found. This is the motivation of importance sampling technique.

The results obtained from Monte Carlo simulation can be improved by variance reduction techniques. Thus, Section 2.5 introduces various variance reduction techniques.

2.5 VARIANCE REDUCTION TECHNIQUES

Variance reduction techniques are meant to improve the results of simulation from various methods by increasing the precision through reducing the variance, although some variance reduction techniques may increase the computational cost. But, on the contrary, these techniques increase the statistical efficiency. However, the choice and application of variance reduction techniques depend on the type of problem under consideration. This section discusses three main types of variance reduction techniques:

1. Common random numbers (CRNs)
2. Antithetic variates
3. Control variates

Each of the above techniques is discussed in detail in Sections 2.5.1 through 2.5.3.

2.5.1 COMMON RANDOM NUMBERS

The basic idea for CRNs is that we compare alternate configurations with similar experimental conditions. Naturally, this technique is suitable for comparing alternate systems rather than analyzing a single system. Let us consider two systems with output parameters X_{1j} and X_{2j}, respectively, for replication j. Suppose that the expected output measure for system i is $\mu_i = E[X_i]$. Since we compare two systems, we intend to find out $d = \mu_1 - \mu_2$ in this case. Now, if $Z_j = X_{1j} - X_{2j}$, then $E[Z_j] = \mu_1 - \mu_2$, that is,

$$\overline{Z}_n = \frac{\sum_{j=1}^{n} Z_j}{n}$$

is an unbiased estimator of d. The variance of $\overline{Z}_n$ can be given by the following:

$$\text{Var}[\overline{Z}_n] = \frac{\text{Var}(Z_j)}{n} = \frac{\text{Var}(X_{1j}) + \text{Var}(X_{2j}) - 2\text{Cov}(X_{1j}, X_{2j})}{n} \tag{2.12}$$

If the systems are simulated separately with different random numbers, then X_{1j} and X_{2j} will be independent, which means that $\text{Cov}(X_{1j}, X_{2j}) = 0$. Now, if we simulate the two systems in such a way that X_{1j} and X_{2j} are positively correlated, then we will have $\text{Cov}(X_{1j}, X_{2j}) > 0$ and our variance will automatically be reduced. In CRN, we try to get this positive correlation by applying the same set of random numbers for all systems.

To implement the CRN, we need to synchronize the random numbers for all the systems that are required to be compared. For example, if we use a random number for generating the interarrival times of one system, then we need to use the same random number for generating the interarrival times for other systems too.

2.5.2 Antithetic Variates

Antithetic variates may be used for a single system. The basic idea is to make pairs of runs for the model so that the small observation of one run is offset by the large observation of the other run. Like CRN, we try to create a positive correlation between several runs. But in the case of an antithetic variate, we need to find out the negative correlation, so that the two observations are negatively correlated. We then find the average of the two observations as a basic data point for analysis. If we use this average, then it will be closer to the common expectation μ of an observation. Thus, antithetic variate tends to create a negative correlation by using complementary random numbers from pairs of runs. If $U(0,1)$ is used for a particular purpose of one run, then $1 - U$ is used for the same purpose of other run, since we need to use complementary random numbers.

Suppose we run n pairs of runs for the simulation model with the resulting observations as $[X_1^{(1)}, X_1^{(2)}], [X_2^{(1)}, X_2^{(2)}], \ldots, [X_n^{(1)}, X_n^{(2)}]$, where $X_j^{(1)}$ is the result from the first run of the jth pair and $X_j^{(2)}$ is the result from the second run of the jth pair, satisfying $E[X_j^{(1)}] = E[X_j^{(2)}] = \mu$, and each pair is independent of the other pairs. Thus, the total number of replications is $2n$.

Next, we find the average of each pair as $X_j = [X_j^{(1)} + X_j^{(2)}]/2$ satisfying $E[X_j^{(1)}] = E(X_j) = E(\overline{X}_n) = \mu$.

As before, the variance can be calculated as follows:

$$\text{Var}(\overline{X}_n) = \frac{\text{Var}(X_j)}{n} = \frac{\text{Var}[X_j^{(1)}] + \text{Var}[X_j^{(2)}] + 2\text{Cov}[X_j^{(1)}, X_j^{(2)}]}{n} \tag{2.13}$$

If the pair of runs is made independent, then we will have $\text{Cov}[X_j^{(1)}, X_j^{(2)}] = 0$, but we want to make it negative, that is, our intention is to make it $\text{Cov}[X_j^{(1)}, X_j^{(2)}] < 0$ so as to reduce the variance as a whole. For an effective antithetic variate, its response to a random number for a particular purpose needs to be monotonic in either direction.

2.5.3 CONTROL VARIATES

The control variate concentrates on the correlation of certain random variables to reduce the variance. Suppose X and Y are two variables in a simulation and we know $\mu_x = E(X)$ and $\mu_y = E(Y)$. If X and Y are positively correlated, it is likely that if $Y > \mu_y$, then $X > \mu_x$. In such a case, we will adjust X downward by some amount. On the contrary, if $Y < \mu_y$, then $X < \mu_x$, and in such a case, we adjust X upward by some amount. Thus, X is adjusted depending on Y. Due to this, Y is said to be a control variate of X. If X and Y are negatively correlated, then the control variate concept will also work.

Now, the important issue comes out to be the amount of adjustment to be done downward or upward. This amount may be represented by $Y - \mu_y$. We use a controlled estimator given by the following:

$$X_c = X - a(Y - \mu_y) \tag{2.14}$$

where:

a is a constant that has the same sign as the sign of correlation between X and Y

For any real number a, $\mu_x = E(X_c)$, since $\mu_x = E(X)$ and $\mu_y = E(Y)$. The variance can be given by the following:

$$\text{Var}(X_c) = \text{Var}(X) + a^2\text{Var}(Y) - 2a\text{Cov}(X,Y) \tag{2.15}$$

Thus, the variance will be reduced if $2a\text{Cov}(X,Y) > a^2\text{Var}(Y)$. Therefore, the value of a should be chosen carefully so as to satisfy this condition. The optimal value of a is as follows:

$$a^* = \frac{\text{Cov}(X,Y)}{\text{Var}(Y)} \tag{2.16}$$

2.6   WHEN TO USE MONTE CARLO SIMULATION

This section shows the kind of situations in which Monte Carlo simulation may be applied. These situations are delineated in the following points:

1. The first and foremost point is that Monte Carlo simulation is applicable to those problems that are difficult to be solved by a simple analytical method. Thus, the complexity of the problem determines the applicability of Monte Carlo simulation.
2. In many cases, although there are alternate methods available, Monte Carlo simulation may be the better and easiest method. For example, complex

repair policies may be dealt with the Monte Carlo simulation rather than an analytical model.

3. Analytical models may be more accurate but may involve assumptions that may make the problem unrealistic. Such assumptions may be included in the problem if it is implemented by Monte Carlo simulation. However, note that the repeated execution of Monte Carlo simulation may result in longer execution times.

4. If modifications are to be incorporated more than once or frequently, then it becomes very difficult and tedious to make the adjustments in an analytical model, whereas in a Monte Carlo simulation model, it may not be difficult to make the adjustments, especially in the case of computerized Monte Carlo simulation.

2.7 APPLICATIONS OF MONTE CARLO SIMULATION

This section provides a brief overview of the existing successful research applications of Monte Carlo simulation in various fields of study. Monte Carlo simulation finds its application from the medical systems, on the one end, to the manufacturing systems, on the other end. We do not intend to review the entire literature but rather intend to show the variety of applications of Monte Carlo simulation.

Smit and Krishna [3] applied Monte Carlo simulation on the absorption and diffusion of hydrocarbons in zeolites. They determined the thermodynamic properties with the help of kinetic Monte Carlo simulation and configurational-bias Monte Carlo simulation. Nilsson et al. [4] used Monte Carlo simulation on photon absorption in X-ray pixel detectors.

Similarly, Monte Carlo simulation has found applications in microwave devices [5], crystal precipitation [6], risk management [7], thermometer calibration [8], pollution control [9], metal–oxide–semiconductor field-effect transistor (MOSFET) application [10], distributed computer systems [11], gamma-ray interaction [12], and so on.

2.8 ADVANTAGES AND DISADVANTAGES OF MONTE CARLO SIMULATION

The advantages and disadvantages of Monte Carlo simulation can be easily evident from Sections 2.3 through 2.6. However, Sections 2.8.1 and 2.8.2 list the advantages and disadvantages gathered from the existing literature, respectively.

2.8.1 ADVANTAGES

The advantages of Monte Carlo simulation can be described through the following points:

1. It is computationally very simple. It does not require an in-depth knowledge of the solution form or the analytical properties.
2. It is easy to be implemented with a computer.

3. The amount of effort that is required to get results of acceptable precision is independent of the dimensions of the variables.
4. The function that can be integrated by Monte Carlo simulation does not need any boundary conditions.

2.8.2 DISADVANTAGES

The disadvantages of Monte Carlo simulation are depicted by the following points:

1. Monte Carlo simulation, an effective simulation technique, is a slow process, since a large number of samples are required to get an acceptable precision of the results.
2. The error in Monte Carlo simulation may be dependent on the distribution of the underlying variable.
3. There is no hard bound on the error of the computed result.
4. With some types of analytical approximation, one can study the behavior of the solution if the initial parameters are changed. This is generally hard to do with Monte Carlo simulation.

2.9 CONCLUSION

This chapter has presented various aspects of Monte Carlo simulation. Starting from its brief history in the introductory section to the advantages and disadvantages, all the possible issues related to Monte Carlo simulation have been discussed in this chapter. In Section 2.1, we present two numerical examples to make the concept easy to understand based on the view that the concepts can better be understood by examples than by theory first. However, starting from such an easy approach, we have presented more in-depth concepts such as types of Monte Carlo simulations and various variance reduction techniques. In addition, Section 2.7 shows the variety of applications of Monte Carlo simulation as evident from the existing literature. Finally, Section 2.8 shows the list of the advantages and disadvantages of Monte Carlo simulation.

REFERENCES

1. Baeurle, S. A. (2009). Multiscale modeling of polymer materials using field-theoretic methodologies: A survey about recent developments. *Journal of Mathematical Chemistry* 46(2): 363–426.
2. Blais, J. A. R. and Zhang, Z. (2011). Exploring pseudo- and chaotic random Monte Carlo simulations. *Computers & Geosciences* 37: 928–934.
3. Smit, B. and Krishna, R. (2001). Monte Carlo simulations in zeolites. *Current Opinion in Solid State & Materials Science* 5: 455–461.
4. Nilsson, H.-E., Dubaric, E., Hjelm, M., and Englund, U. (2003). Monte Carlo simulation of the transient response of single photon absorption in X-ray pixel detectors. *Mathematics and Computers in Simulation* 62: 471–478.
5. Ravaioli, U., Lee, C. H., and Patil, M. B. (1996). Monte Carlo simulation of microwave devices. *Mathematical and Computer Modelling* 23(8/9): 167–179.

6. Falope, G. O., Jones, A. G., and Zauner, R. (2001). On modelling continuous agglomerative crystal precipitation via Monte Carlo simulation. *Chemical Engineering Science* 56: 2567–2574.

7. Rezaie, K., Amalnik, M. S., Gereie, A., Ostadi, B., and Shakhseniaee, M. (2007). Using extended Monte Carlo simulation method for the improvement of risk management: Consideration of relationships between uncertainties. *Applied Mathematics and Computation* 190: 1492–1501.

8. Shahanaghi, K. and Nakhjiri, P. (2010). A new optimized uncertainty evaluation applied to the Monte-Carlo simulation in platinum resistance thermometer calibration. *Measurement* 43: 901–911.

9. Vallée, F., Versèle, C., Lobry, J., and Moiny, F. (2013). Non-sequential Monte Carlo simulation tool in order to minimize gaseous pollutants emissions in presence of fluctuating wind power. *Renewable Energy* 50: 317–324.

10. Bournel, A., Aubry-Fortuna, V., Saint-Martin, J., and Dollfus, P. (2007). Device performance and optimization of decananometer long double gate MOSFET by Monte Carlo simulation. *Solid-State Electronics* 51: 543–550.

11. Camarasu-Pop, S., Glatard, T., da Silva, R. F., Gueth, P., Sarrut, D., and Benoit-Cattin, H. (2013). Monte Carlo simulation on heterogeneous distributed systems: A computing framework with parallel merging and checkpointing strategies. *Future Generation Computer Systems* 29: 728–738.

12. Kamboj, S. and Kahn, B. (2003). Use of Monte Carlo simulation to examine gamma-ray interactions in germanium detectors. *Radiation Measurements* 37: 1–8.

3 Introduction to Probability Theory

3.1 INTRODUCTION

The literal meaning of probability is chance. The theory of probability deals with the laws of chances of occurrences of various phenomena that are unpredictable in nature. Probability has its origin in the game of chance. In the early sixteenth and seventeenth centuries, the owners of the gambling houses in Europe intended to explore whether one could find out the probability of various events that take place during the gambling game, such as tossing a die or a coin, and roulette wheels. So, they contacted some of the prominent mathematicians of that time, say Pascal and Fermat. Thus, the theory of probability started its journey in the sixteenth century.

Games of chance were very popular in both Greece and Rome. But the number system followed by the Greeks was not supportive of the algebraic calculations, and for this reason scientific development on games of chance could not take place there at that time. During the second half of the first millennium, the Hindus and the Arabs started developing modern arithmetic systems. From there, the scientific ideas about the number systems started. Later some time during the sixteenth century, an Italian mathematician named Girolamo Cardano proposed a method of calculating probabilities for games of chance with dice and cards. He proposed his ideas in his book. After that, scientists Fermat and Pascal, during the seventeenth century, raised several questions related to the theory of probability that motivated the study of the subject further.

During the eighteenth century, Jacon Bernoulli introduced the law of large numbers and further instigated the thought through his coin tossing experiments. His study made a breakthrough in the area that indulged the ideas in the minds of several mathematicians such as Daniel Bernoulli, Leibnitz, Bayes, and Lagrange. Their significant contributions further enriched the field of probability theory. Later, De Moivre proposed a normal distribution and central limit theorem, which opened a new era of study in the respective field of study.

After this, during the nineteenth century, Laplace introduced his own concepts and described the importance of probability theory in numerical study. He also introduced some more versions of the central limit theorem. Later, Gauss and Lgendre applied probability concepts to the astronomical predictions, and Poisson introduced the concept known as the Poisson distribution. Chebyshev and his students Markov and Lyapunov studied the limit theorem and made significant contributions to probability theory. The twentieth century saw the emergence of axiomatic definitions of probability, an evolution from the relative frequency definition of probability.

Since then, probability theory has emerged as an important and essential theory for the uncertain events in nature.

One of the features of probability is that the phenomena which we are interested in are random in nature. So, for example, if we consider tossing a die, then we do not know which face numbered 1, 2, 3, 4, 5, and 6 will appear. But in the long run, if we toss the die many times, the proportion of the number of occurrences of, say 6, will be 1/6. Similarly, in the tossing of a coin, we do not know whether we will get a head or a tail in a single toss. But if we toss for a large number of times, then we will know that there will be a head approximately 50% of the times and there will be a tail approximately 50% of the times.

This long-term behavior is known as "statistical regularity," which encourages us to study probability. One can get a similar kind of observation in experiments that are connected in real life, such as experiments in physics and genetics or practically any phenomenon in real life. While promoting a new policy, an insurance company would like to know how many of the people will survive up to the age of maturity. For example, if the policy matures at the age of 60, then they would like to know the chance that they will reach age 60 and therefore may get the benefits that are due to them. For each individual person, it is not possible to know whether he will die at the age of 60, but in the population as a whole, it is possible to predict the percentage of people dying beyond 60. A similar kind of statistical regularity is observed and used in weather predictions, crop growth predictions, economic growth, the financial situation of a country, and so on. Hence, in most of these cases, although things may look like they are predesigned and predetermined, but actually, there will be several conditions that regulate their occurrences of the final phenomenon, and therefore, one can treat them as a random phenomenon. Before going into the depth of probability theory, let us get acquainted with some related terms and concepts provided in Section 3.2.

The remaining sections of this chapter are arranged as follows: Section 3.2 provides some ideas and definitions that are required for the study of probability theory; Section 3.3 provides a brief introduction to set theory that is required for the concept of probability theory; Section 3.4 mentions some counting techniques that are required for the calculation of probability; Section 3.5 defines the probability concept; Section 3.6 provides some numerical examples to explain how to calculate simple probability values; Section 3.7 describes some laws of probability; and Section 3.8 concludes this chapter.

3.2 DEFINITIONS RELATED TO PROBABILITY THEORY

The definitions provided in this section can help in understanding probability rules and concepts. We provide a list of definitions relevant to the study of probability theory as follows:

1. *Experiment.* An experiment means the observing of something happening or conducting something under certain conditions, which results in some outcome. Consider the rainfall phenomenon. Rainfall is the consequence of several

factors such as cloud formation, some alien occurrences, and humidity issues. There are various other factors that lead to rainfall. Observing the weather is an experiment. Similarly, for example, we may want to find the yield of a particular crop in a particular field. The yield of a crop depends on several factors, such as the type of seeds, the plot of land where it is cultivated, the irrigation procedure, and some mechanical procedures that are used in farming. So, the entire process is a random experiment. The outcome is recorded as the final yield for the crop.

There are broadly two types of experiments: (1) deterministic experiments and (2) random experiments. However, we are concerned more with random experiments.

a. Deterministic experiment—If an experiment is conducted under certain conditions, then it results in known outcomes. Such experiments are called deterministic experiments.

b. Random experiment—In this experiment, although we may fix the conditions under which the trials or the experiments are conducted, the outcome is still uncertain. This is the reason for calling the experiment a random experiment.

Consider the following examples. All of these examples depict random experiments.

   i. Tossing of a coin: Although we can fix a lot of conditions for a coin tossing experiment, the outcome is still unknown. The conditions may include the kind of coin that is in our hand and the way we are holding the coin. But even then, when we toss the coin and it falls, the outcome is uncertain—head or tail, or in certain situations, it may stand on its side.

   ii. Tossing of a die: Again the conditions are similar. But after tossing, which face will appear is quite unknown.

   iii. Drawing a card from a deck of cards

   iv. Birth of a child

   v. Amount of rainfall during monsoon season in a geographical area

   vi. Yield of a crop of a certain food grain in a state

   vii. The time taken to complete a 100-m race by an athlete

   viii. Life of a bulb

   ix. Life of a mechanical instrument: Even if they are made by the same process, the life is still uncertain.

   x. Number of defective items produced by a company

In all these phenomena, the conditions of the experiments are fixed. For example, when we want to find the time taken by an athlete to complete a 100-m race, the conditions are fixed. For example, the ground is fixed, the starting time is fixed, the athlete is in perfect health, the person who will direct the race is prepared, and the person who will record the time is prepared. However, how much actual time the athlete is going to take to complete a 100-m race will always be uncertain. For example, it may be 10, 9.7, or 10.1 seconds.

All the above experiments are random experiments, and in the subject of probability, we are concerned only with the random experiments.

2. *Sample space.* The set of all possible outcomes of a random experiment is known as the sample space. The sample space is usually denoted by the symbol Ω or S. Examples may include the following:
   a. Tossing of a coin: $\Omega = \{H,T\}$, where H is head and T is tail
   b. Tossing of a die: $\Omega = \{1,2,3,4,5,6\}$
   c. Drawing a card from a deck of cards: $\Omega = \{D_1,D_2,...,D_{13},C_1,C_2,...,$ $C_{13},H_1,H_2,...,H_{13},S_1,S_2,...,S_{13}\}$, where D is diamond, H is heart, C is club, and S is spade
   d. Birth of a child: $\Omega_1 = \{M,F\}$, where M is male and F is female; $\Omega_2 = \{H,U\}$, where H is healthy and U is unhealthy

3. *Event.* An event is any subset of the sample space. A head or tail is an event. Events can be classified into two types as follows:
   a. Impossible event—Since every event is a subset of a sample space, the empty set $\Phi \in \Omega$ corresponds to an impossible event.
   b. Sure event—Since Ω is a subset of Ω, Ω is a sure event. For example, when tossing a die, if one says that 7 occurs, then it is an impossible event since the number 7 is not a subset of the sample space.

4. *Mutually exclusive events.* Two events can be mutually exclusive if the occurrence of any of the events prevents the occurrence of the other. For example, in a coin tossing experiment, if a head appears, then a tail cannot appear. Thus, the occurrence of the events, head and tail, is mutually exclusive.

However, to understand the theory of probability, the background knowledge of set theory is required.

3.3   BRIEF INTRODUCTION TO SET THEORY

A set can be defined as a group of objects put together [1]. For example, a group of students can represent a set. Some of the examples of sets are given as follows:

- A set of positive even numbers less than 10, $N = \{0,2,4,6,8\}$
- A set of vowels, $V = \{a,e,i,o,u\}$
- A set representing six students in a study group, *names* = {*Paul,Mertin, John,Maria,Tom*}
- A set of basic shapes, $S = \{square,rectangle,circle,triangle,pentagon, hexagon\}$
- A set of European countries, $C = \{England,France,Germany,Poland,Italy, Norway\}$
- A set of selected fruits, $F = \{mango,coconut,palm,banana,apple\}$
- A set of symbols, $S = \{\alpha,\chi,\beta,\delta,\phi\}$
- A set of marks in mathematics obtained by a group of students, $M = \{88,100,40,75,80,62,55,96\}$
- A set of grades, $G = \{O,S,A,B,C,F\}$

Generally, a set contains similar elements, but there are sets that can also contain dissimilar elements. Some examples of sets having dissimilar elements are given as follows:

- Information for a particular student, $I = \{JU1A6378, Xiang, 24, China\}$, which means that a student named "Xiang" with identification number "$JU1A6378$" of age 24 years is from China.
- A set of elements to constitute a variable name, $S = \{c, t, 2, 1, \_, 1, 4\}$

The members of a set are known as elements. Two sets are said to be equal if they have the same elements. Sets can be represented in two ways: (1) to mention the elements with a pair of second braces, $S = \{\ldots, \ldots, \ldots\}$ and (2) to use set builder notation. Some examples of set builder notation are given as follows:

- $S = \{x | 1 \leq x \leq 10\}$, which means that x is a set of values lying between 1 and 10.
- $S = \{x | x \neq 0 \text{ and } -2 \leq x \leq 4\}$, which means that the value of x will be any value between -2 and 4 except the value 0.
- $S = \{x | x \text{ is a real number}\}$.

Although a detailed introduction to sets is feasible here, some of the concepts that are compulsorily required are as follows:

- Two sets A and B are equal if and only if A contains every element of B and B contains every element of A. Mathematically, A and B are equal if, for all x, $x \in A$ and $x \in B$. In other words, it can be said that the two sets A and B are equal if A is a subset of B and B is a subset of A.
- Let A and B be two sets. B is a subset of A if and only if every element of B is an element of A. Mathematically, $B \subset A$ if and only if A and B are two subsets under the universal set U and for all x, $x \in B \Rightarrow x \in A$.
- For every set A, $A \in A$.
- Transitivity property of sets: If $A \in B$ and $B \in C$, then according to the transitivity property, we have $A \in C$.
- In a universal set, there is one and only set that has no element. This set is called the null set. An empty set of a null set is denoted by the symbol Φ (phi).
- For any set A, $\Phi \in A$.
- There are various types of set operations. Some of them are union (denoted by the symbol "$\cup$"), intersection (denoted by the symbol "$\cap$"), subtraction (denoted by the symbol "$-$"), and complement (denoted by the symbol "c"); for example, the union of two events A and B is given by $A \cup B$, which indicates the occurrence of at least one of the events A and B. $\bigcup_{i=1}^{n} A_i$: Occurrence of at least one of A_i events, $i = 1, 2, \ldots$
- The intersection of two events is denoted by $A \cap B$, which indicates a simultaneous occurrence of events A and B. $\bigcap_{i=1}^{n} A_i$: a simultaneous occurrence of all A_i events. We can have $\bigcup_{i=1}^{n} A_i = \Omega$, which means that all the points in Ω are contained in $\bigcup_{i=1}^{n} A_i$. Such events are called exhaustive events.

- For any set A, $A \cup \Phi = A$ and $A \cap \Phi = \Phi$.
- If $A \cap B = \Phi$, events A and B cannot occur together. They are called mutually exclusive events, that is, the happening of one of these events excludes the possibility of the occurrence of the other event. If we have a collection of events, $A_1, A_2, ...,$ such that $A_i \cap A_j = \Omega$ when $i \neq j$, then we say that $A_1, A_2, ...$ are pairwise disjoint or mutually exclusive events.
- The symbol A^c means "not a happening of event A."
- The symbol $A - B$ means the "occurrence of event A but not the occurrence of event B":

$$A - B = A \cap B^c$$

Another basic concept of set theory is the power set [2]. The power set is the set of all possible subsets of a given set. For example, consider the set $A = \{2,4,6\}$. The power set of set A is given by the following:

$$P = \{\Phi, \{2\}, \{4\}, \{6\}, \{2,4\}, \{2,6\}, \{4,6\}, \{2,4,6\}\} \tag{3.1}$$

We may call the events $A_1, A_2, ...$ equally likely events. Events A and B are equally likely events if A and B have the same chances of appearing in a random experiment. The above given terminologies are some of the terminologies occurring in reality.

3.3.1 VENN DIAGRAM

A Venn diagram is a pictorial way to express the relationships among the sets. In a Venn diagram, a set is represented by a circle. Thus, if A and B are two sets, they will be represented by two circles as shown in Figure 3.1.

Figure 3.1 shows two circles representing the two sets A and B separately. The same diagram may be used to show the two disjoint sets A and B. Two sets A and B are said to be disjoint if the two sets are completely dissimilar and unequal. If B is a subset of A, then Figure 3.2 will be the representative figure for the relation between the sets A and B.

The intersection of the two sets A and B represents the common elements between the two sets and is represented by Figure 3.3. The shaded region represents $A \cap B$. For example, suppose the sets A and B are given by $A = \{1,2,3,4,5,6\}$ and $B = \{2,4,6,8,10\}$; then their intersection will be given by the following:

$$A \cap B = \{2,4,6\} \tag{3.2}$$

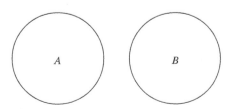

FIGURE 3.1   Two disjoint sets A and B.

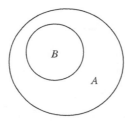

FIGURE 3.2   Venn diagram representing the set $B \subset A$.

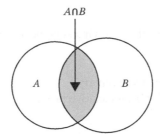

FIGURE 3.3   Intersection of the two sets A and B.

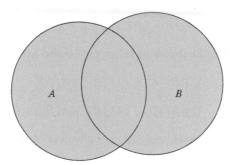

FIGURE 3.4   Union of the two sets A and B.

The union of the two sets A and B represents those elements that belong to both sets A and B. Mathematically, the union $A \cup B$ implies $\{x | x \in A \text{ or } x \in B\}$. The union of the two sets A and B is represented by Venn diagram in Figure 3.4. The shaded region represents $A \cup B$.

For the three sets A, B, and C, the following statements are true:

- $A \cap A = A$
- $A \cap B = B \cap A$ (commutative property)
- $(A \cap B) \cap C = A \cap (B \cap C)$ (associative property)
- $A \cap B \subset A$ and $A \cap B \subset B$
- $A \cap \Phi = \Phi$

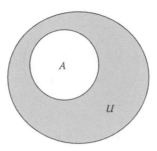

FIGURE 3.5   Venn diagram for the complement of A.

For the three sets A, B, and C, the following statements on union are true:

- $A \cup A = A$
- $A \cup B = B \cup A$ (commutative property)
- $(A \cup B) \cup C = A \cup (B \cup C)$ (associative property)
- $A \subset A \cup B$ and $B \subset A \cup B$
- $A \cup \Phi = A$

From the previous example where the sets A and B are given by $A = \{1,2,3,4,5,6\}$ and $B = \{2,4,6,8,10\}$, the union of the sets A and B is given by the following:

$$A \cup B = \{1,2,3,4,5,6,8,10\} \tag{3.3}$$

For the two sets A and B, the following relations are also true:

$$A \cap (B \cup C) = (A \cap B) \cup (A \cap C) \tag{3.4}$$

$$A \cup (B \cap C) = (A \cup B) \cap (A \cup C) \tag{3.5}$$

If A is a subset of the universal set U, the complement of the set A is given by the following:

$$A^c = U - A \tag{3.6}$$

The complement can be expressed in another way as follows:

$$A^c = \{x \mid x \in U \text{ and } x \notin A\} \tag{3.7}$$

The complement A^c can be given by the Venn diagram in Figure 3.5. The shaded region represents $A^c = U - A$.

Venn diagrams can be used to represent various complex set operations. Some of the examples are given as follows:

Example 1

The set expression $A \cap B \cap C$ can be shown by the Venn diagram in Figure 3.6.

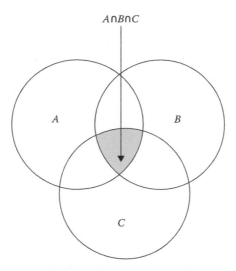

FIGURE 3.6 Venn diagram representing the set $A \cap B \cap C$.

Example 2

The set expression $A \cup B \cup C$ can be shown by the Venn diagram in Figure 3.7.

Example 3

The set expression $(A \cup B) - C$ can be shown by the Venn diagram in Figure 3.8. To draw such an expression, first $A \cup B$ is drawn as shown by the shaded region in Figure 3.8a, then the region of C is identified (Figure 3.8b), and finally the region of C is excluded from $A \cup B$ to draw $(A \cup B) - C$ (Figure 3.8c).

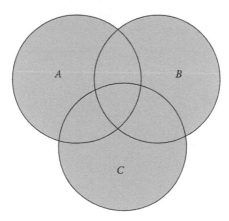

FIGURE 3.7 Venn diagram representing the set $A \cup B \cup C$.

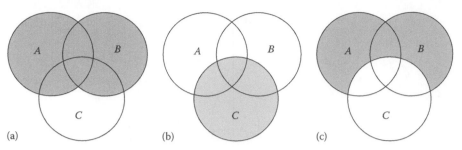

(a)   (b)   (c)

FIGURE 3.8 Venn diagram representing the set $(A \cup B) - C$.

Example 4

The set expression $A \cup (B \cap C^c)$ can be shown by the Venn diagram in Figure 3.9. To draw this diagram, first C is drawn as shown in Figure 3.9a, followed by the representation of C^c in Figure 3.9b. Figure 3.9c represents the set $B \cap C^c$, followed by the final expression $A \cup (B \cap C^c)$ (Figure 3.9d) obtained by the union of the set $B \cap C^c$ and the set A.

The Venn diagram is a friendly way to understand and represent the favorable event while calculating the probability values. Before starting the concepts of probability theories, a brief introduction to various counting techniques is required. Thus, Section 3.4 provides a brief overview of the counting techniques that will be required in the subsequent sections.

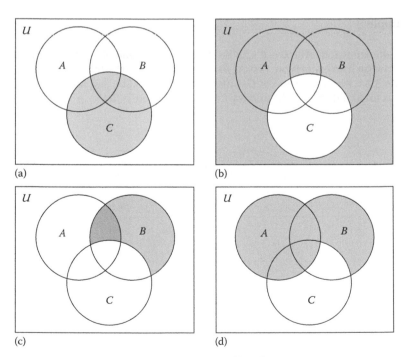

FIGURE 3.9 Venn diagram representing the set $A \cup (B \cap C^c)$.

3.4   COUNTING TECHNIQUES

Counting techniques are mathematical techniques that are used to find the number of outcomes in classical and other related probability concepts. These concepts are listed as follows:

1. *Basic counting techniques.* If several processes occur in such a way that process A occurs in p ways, process B occurs in q ways, and process C occurs in r ways, then all the processes can be performed in $p \times q \times r \cdots$ ways.

2. *Permutation.* The total number of ways in which n objects can be arranged is given by

$$^{n}P_{r} = n(n-1)(n-2)...(n-r+1) \qquad (3.8)$$

3. *Permutation with repetition.* The total number of ways in which n objects can be arranged, where p objects are alike, q objects are alike, r objects are alike, and so on, is given by

$$\frac{n!}{p!\,q!\,r!\,...}$$

4. *Combination.* The total number of possible groups formed by taking r objects out of n different objects is given by

$$^{n}C_{r} = \frac{n!}{r!(n-r)!} = \frac{n(n-1)(n-2)...(n-r+1)}{r!} \qquad (3.9)$$

5. *Combination of any number of objects at a time.* The total number of possible groups formed by taking any number of objects is given by

$$^{n}C_{1} + {}^{n}C_{2} + {}^{n}C_{3} + \cdots + {}^{n}C_{n}$$

6. *Arranging objects in line or circle.* The total number of ways in which n objects can be arranged in a line is given by

$$n! = 1 \times 2 \times \cdots \times n$$

The total number of ways in which n objects can be arranged in a circle is given by

$$(n-1)!$$

7. *Choosing balls from an urn.* The total number of ways of choosing a number of white balls, b number of black balls, c number of red balls, and so on from an urn containing A number of white balls, B number of black balls, C number of red balls, and so on is given by

$$^{A}C_{a}.{}^{B}C_{b}.{}^{C}C_{c}...$$

8. *Ordered partitions with distinct objects.* The total number of ways in which n distinct objects can be distributed into r partitions is given by r^{n}

If we want the partitions to contain exactly $n_1, n_2, n_3, \ldots$ number of objects, then the total number of ways in which n objects can be partitioned into r partitions is given by

$$\frac{n!}{n_1! n_2! n_3! \ldots n_r!}$$

9. *Ordered partitions with identical objects.* The total number of ways in which n identical objects can be distributed into r partitions is given by

$$^{n+r-1}C_{r-1}$$

The total number of ways in which n identical objects can be distributed into r partitions such that no partition stays empty is given by

$$^{n-1}C_{r-1}$$

10. *Sum of points on dice.* If a total of n dice are thrown, then the total number of ways in which a total of r points is obtained is given by the coefficient of x^r in the expression given by

$$(x + x^2 + x^3 + x^4 + x^5 + x^6)^n$$

11. *Derangements and matches.* If n number of objects numbered as $1,2,3,\ldots$ are placed at random in n places, also numbered as $1,2,3,\ldots$, then a match is said to occur if an object of a particular number is placed on the location of the same number. The possible number of ways or permutations in which no match occurs can be given by

$$n! \left\{ 1 - \frac{1}{1!} + \frac{1}{2!} - \frac{1}{3!} + \cdots + (-1)^n \frac{1}{n!} \right\}$$

The number of ways or permutations in which exactly r matches occur is given by

$$\frac{n!}{r!} \left\{ 1 - \frac{1}{1!} + \frac{1}{2!} - \frac{1}{3!} + \cdots + \frac{(-1)^{n-r}}{(n-r)!} \right\}$$

3.5   DEFINITION OF PROBABILITY

In this section, probability can be defined in three different aspects.

- Classical definition of probability
- Relative frequency definition of probability
- Axiomatic definition of probability

Definitions of each of these aspects are provided in Sections 3.5.1 through 3.5.3.

3.5.1 CLASSICAL DEFINITION OF PROBABILITY

Suppose a random experiment has N possible outcomes that are mutually exclusive, exhaustive, and equally likely. Suppose M of these outcomes is favorable to the occurrence of an event A. In such a case, the probability of the occurrence of event A is given by

$$P(A) = \frac{M}{N} \qquad (3.10)$$

Although this definition is very effective for finding the simple probabilities of events, this definition has certain drawbacks, which are as follows:

- N does not need to be finite. But this definition considers finite N only.
- The definition is circular in nature as it uses the term "equally likely," which means outcomes with equal probability. But this may not be the reality.

3.5.2 RELATIVE FREQUENCY DEFINITION OF PROBABILITY

Suppose a random experiment is conducted a large number of times independently under identical conditions. Let a_n denote the number of times event A occurs in n trials of an experiment. Then, we define the probability of the occurrence of event A as follows:

$$P(A) = \frac{\lim}{n \to \infty} \frac{a_n}{n}, \text{provided the limit exists} \qquad (3.11)$$

Let us consider the trial of conducting a coin toss. Suppose we have conducted n number of trials. Thus, the outcomes of each of these n trials may be either head (H) or tail (T). Thus, the n trials may look like the following sequence:

$$HHTHHTHHT...$$

Thus, expression (3.4) may be calculated as follows:

$$\frac{a_n}{n} \to \frac{1}{1} \text{ (after first trial)}$$

$$\frac{a_n}{n} \to \frac{2}{2} \text{ (after second trial)}$$

$$\frac{a_n}{n} \to \frac{2}{3} \text{ (after third trial)}$$

$$\frac{a_n}{n} \to \frac{3}{4} \text{ (after fourth trial)}$$

$$\frac{a_n}{n} \to \frac{4}{5} \text{ (after fifth trial),...}$$

The relative frequency definition is the most effective since it is based on the actual experience and what the subject of probability should be all about. The drawbacks of the relative frequency definition are as follows:

- Many times, the observation may be complex or too costly, so we may not be able to observe in those cases.
- We are assuming that the experiment is being observed. But actual observation of the experiment may not be possible sometimes.

3.5.3 AXIOMATIC DEFINITION OF PROBABILITY

An axiomatic definition of probability was given by Kolmogorov in 1933. Let (Ω, B) be a measurable space. A set function $P:B \rightarrow R$ is said to be a probability function if it satisfies the following axioms, where (Ω, B, P) is called the probability space.

- $C_1: P(\Omega) = 0$
  Proof: Let us take $A_1 = \Omega$, $A_2 = A_3 = \cdots = \Phi$ in P_3.
    Then, $P(\Omega) = P(\Omega) + P(\Phi) + P(\Phi) + \ldots$.
    Since $P(\Omega) = 1$ and it implies $P(\Phi) \geq 0$ [since $P(A) \geq 0$], we must have
      $P(\Phi) = 0$.
- C_2: For any finite collection $A_1, A_2, ..., A_n$ of pairwise disjoint sets in B, we have the following finite additivity expression:

$$P\left(\bigcup_{i=1}^{n} A_i\right) = \sum_{i=1}^{n} P(A_i) \tag{3.12}$$

- C_3: If P is a monotone function, that is, $A \subset B$, then $P(A) \leq P(B)$.
- C_4: For any $A \in B$, $0 \leq P(A) \leq 1$.
- C_5: $P(A^c) = 1 - P(A)$.

Some of the further consequences of the axiomatic definition are outlined as follows:

- For any events A and B,

$$P(A \cup B) = P(A) + P(B) - P(A \cap B) \tag{3.13}$$

which, in generalized form, can be expressed as follows:

$$P\left(\bigcup_{i=1}^{n} A_i\right) = \sum_{i=1}^{n} P(A_i) - \sum \sum_{i<j} P(A_i \cap A_j)$$

$$+ \sum \sum \sum_{i<j<k} P(A_i \cap A_j \cap A_k) - \cdots + (-1)^{n+1} P\left(\bigcap_{i=1}^{n} A_i\right) \tag{3.14}$$

- If $\{A_n\}$ is a monotonic sequence of sets in B, then

$$P\left(\lim_{n \to \infty} A_n\right) = \lim_{n \to \infty} P(A_n) \tag{3.15}$$

- We know that $P(A \cup B) = P(A) + P(B) - P(A \cap B)$. Thus, we can write

$$P(A \cup B) \le P(A) + P(B) - P(A \cap B) \tag{3.16}$$

This is called subadditivity.
- Subadditivity probability function: For $A_1, A_2, ..., A_n \in B$, we have

$$P\left(\bigcup_{i=1}^{n} A_i\right) \le \sum_{i=1}^{n} P(A_i) \tag{3.17}$$

- For any countable sequence $\{A_i\} \in B$,

$$P\left(\bigcup_{i=1}^{\infty} A_i\right) \le \sum_{i=1}^{\infty} P(A_i) \tag{3.18}$$

- Bonferroni inequality: For any events $A_1, A_2, ..., A_n \in B$, we have

$$\sum_{i=1}^{n} P(A_i) - \sum\sum_{i<j} P(A_i \cap A_j) \le P\left(\bigcup_{i=1}^{n} A_i\right) \le \sum_{i=1}^{n} P(A_i) \tag{3.19}$$

- Boole's inequality: Let $\{A_n\}$ be a sequence of sets in B; then

$$P\left(\bigcap_{i=1}^{\infty} A_i\right) \ge 1 - \sum_{i=1}^{\infty} P(A_i^c) \tag{3.20}$$

3.6 NUMERICAL EXAMPLES OF CLASSICAL APPROACH TO PROBABILITY

This section provides some simple examples of probability that apply all the counting techniques discussed in Section 3.4. In the classical approach, discussed in Section 3.5.1, all possible outcomes of an experiment are first enumerated. Then, the favorable events among them for which the probability is required are counted. This number is divided by the total possible outcomes. The basic formula for this calculation is provided in expression (3.10). The examples are provided as follows:

Example 5

What is the possibility of getting two heads if two unbiased coins are tossed?

In this example, two unbiased coins are tossed. For each coin, there are two possible events—head or tail. Thus, for the case of tossing two unbiased coins, any of the following events may occur:

HH (both are heads)
HT (the first one is head and the second one is tail)

TH (the first one is tail and the second one is head)
TT (both are tails)

Thus, it is observed that there are a total of four events that can occur. Thus, following expression (3.10), $N = 4$. Among these four occurrences, there is only one event in which a head appears in both trials, denoted by *HH*. Thus, we have $M = 1$. Thus, the required probability is

$$P(A) = \frac{M}{N} = \frac{1}{4}$$

Example 6

An urn contains 10 white balls and 20 red balls. One ball is drawn from the urn. What is the probability that the chosen ball will be red?

The question says that there are 10 white balls and 20 red balls. This means that there are $10 + 20 = 30$ outcomes regarding the chosen ball, that is, the chosen ball can be any of these 30 balls. Thus, the total possible outcomes is $N = 30$.

Now, the chosen ball should be red. Since there are 20 red balls, the chosen ball can be any of these 20 red balls. Thus, the number of favorable outcomes is $M = 20$, and the required probability is as follows:

$$P(A) = \frac{M}{N} = \frac{20}{30} = \frac{2}{3}$$

Example 7

Suppose that there are four children in a family. Find the probability that all these four children have separate birthdays. The question says that there are 360 days in a year.

Assume for the sake of calculation that there are 360 days in a year. Thus, the first child can have a birthday on any of these 360 days and the second child can also have a birthday on any of these 360 days. Similarly, each of the other two children can also have their birthday on any of these 360 days. Thus, the total number of possible ways in which the children have their birthdays is the total number of possible outcomes and is given by the following:

$$N = 360 \times 360 \times 360 \times 360$$

Since the question says that each child will have a separate birthday, the first child can have a birthday on any of these 360 days. Thus, the remaining days will be $360 - 1 = 359$ days. The second child can have a birthday in any of these 359 days. Thus, now the remaining days will be $359 - 1 = 358$ days. The third child will have a birthday on any of these 358 days. Now, the remaining days will be $358 - 1 = 357$ days. The fourth child will have a birthday on any of these 357 days. Thus, the number of favorable outcomes is

$$M = 360 \times 359 \times 358 \times 357$$

The required probability is given by the following:

$$P(A) = \frac{M}{N} = \frac{360 \times 359 \times 358 \times 357}{360 \times 360 \times 360 \times 360} = 0.983$$

Example 8

There are 20 distinct balls. These balls are distributed at random into five boxes. What is the probability that a box will contain exactly three balls?

There are 20 distinct balls and 5 boxes. The first ball may go into any of the five boxes; the second ball can also go into any of the five boxes; the third ball may go into any of the five boxes; and so on. Thus, the 20 balls can be distributed into 5 boxes in 5^{20} different ways, and the total number of possible outcomes is $N = 5^{20}$.

Among these 20 balls, 3 balls will be moved to a particular box. The 3 balls can be chosen from the total of 20 balls in $^{20}C_3$ ways. The remaining number of boxes is $5 - 1 = 4$, and the remaining number of balls is $20 - 3 = 17$. Thus, in the same way as earlier, these 17 balls can be distributed in 4 boxes in 4^{17} different ways. The number of favorable cases is $M = {}^{20}C_3 \times 4^{17}$, and the required probability is as follows:

$$P(A) = \frac{M}{N} = \frac{{}^{20}C_3 \times 4^{17}}{5^{20}} = \frac{1140 \times 4^{17}}{5^{20}} = 0.205$$

3.7   LAWS OF PROBABILITY

In this section, first we discuss conditional probability. The conditional probability is calculated when only the partial outcome of an experiment is available. For example, a fair die is rolled. Suppose B is an event in which an even number occurs and E is an event in which number 2 occurs.

Now since a die has six faces numbered 1, 2, 3, 4, 5, and 6, and there is only one "2" in the sample space, the probability that number 2 occurs is 1/6. Now if we know that an even number has occurred, then the probability of occurrence of number 2 based on the fact that an even number has occurred is $P(E|B\ has\ occurred) = 1/3$.

Thus, let (Ω, β, P) be a probability space, and let the probability of occurrence of event $B \in \beta$ be $P(B) \geq 0$. Then, for any event $A \in \beta$, the conditional probability of A, given that B has already occurred, is as follows:

$$P(A \mid B) = \frac{P(A \cap B)}{P(B)} \qquad (3.21)$$

From expression (3.14), we can write

$$P(A \cap B) = P(B)P(A \mid B) \qquad (3.22)$$

Thus, the probability of simultaneous occurrence of A and B is equal to the conditional probability of A given B, multiplied by $P(B)$ occurring alone. If we interchange the roles, we will get

$$P(A \cap B) = P(A)P(B \mid A) \qquad (3.23)$$

This is known as the multiplication rule.

$$P(A \cap B) = P(A)P(B \mid A)$$

$$= P(B)P(A \mid B) \tag{3.24}$$

The generalized multiplication rule can be depicted in the following expression (3.25). Let $A_1, A_2, ..., A_n \in \beta$, with $P(\cap_{i=1}^{\infty} A_i) > 0$. Then, we can have the following relation:

$$P\left(\bigcap_{i=1}^{\infty} A_i \right) = P(A_1)P(A_2 \mid A_1)P(A_3 \mid A_1 \cap A_2)...P\left(A_n \mid \bigcap_{i=1}^{n-1} A_i \right) \tag{3.25}$$

Next, we provide the theorem on total probability. Let $B_1, B_2, ...$ be pairwise disjoint (i.e., mutually exclusive) events with $B = \cup_{i=1}^{\infty} B_i$. Then, for any event A, we have

$$P(A \cap B) = \sum_{j=1}^{\infty} P(A \mid B_j)P(B_j) \tag{3.26}$$

Furthermore, if $P(B) = 1$, or $B = \Omega$, then we have

$$P(A) = \sum_{j=1}^{\infty} P(A \mid B_j)P(B_j) \tag{3.27}$$

Let us take an example. Suppose a calculator manufacturer purchases his integrated circuits (ICs) from suppliers B_1, B_2, and B_3 with 40% from B_1, 30% from B_2, and 30% from B_3. Suppose 1% of the supply from B_1 is defective, 5% of the supply from B_2 is defective, and 10% of the supply from B_3 is defective. What is the probability that a randomly selected IC from the manufacturer's stock is defective?

Let event A be denoted by the following:

$$A: \text{ IC is defective}$$

Thus, the probability that the IC is defective is given by the following:

$$P(A) = \sum_{j=1}^{3} P(A \mid B_j)P(B_j)$$

$$= P(A \mid B_1)P(B_1) + P(A \mid B_2)P(B_2) + P(A \mid B_3)P(B_3)$$

$$= 0.01 \times 0.4 + 0.05 \times 0.3 + 0.1 \times 0.3$$

$$= 0.004 + 0.015 + 0.03$$

$$= 0.049$$

Sometimes we know the final outcome. Now what is the probability that the outcome was caused by something. This kind of probability is called posterior probability.

Suppose, in the previous example, we know that B_1, B_2, and B_3 are the suppliers, and we take 40%, 30%, and 30% from B_1, B_2, and B_3, respectively. Being defective is the consequence of that. Now someone takes the final product which is found to be defective. What is the probability that the defective item comes from B_1, B_2, or B_3?

The respective theorem is called Bayes' theorem. Suppose $B_1, B_2, \ldots$ are pairwise disjoint events with $\bigcup_{j=1}^{\infty} B_j = \Omega$. This means that they are too exhaustive. We are also given the probabilities $P(B_i) > 0$, $i = 1, 2, \ldots$ Then for any event A, with $P(A) > 0$, we have the following Bayes' formula:

$$P(B_i \mid A) = \frac{P(A \mid B_i)P(B_i)}{\sum_{j=1}^{\infty} P(A \mid B_j)P(B_j)} \qquad (3.28)$$

Thus, Bayes' theorem gives the cause-and-effect relationship among events.

Now, if the occurrence of one event does not influence the occurrence of the other event, then the two events are called independent events. For a pair of independent events, A and B, we have the following relation:

$$P(AB) = P(A)P(B) \qquad (3.29)$$

3.8 CONCLUSION

In this chapter, the basic concepts related to probability have been presented. Three different approaches (classical approach, relative frequency approach, and axiomatic approach) to the definition of probability have been discussed. Numerical approaches on the application of classical approach on various problems have also been provided. To depict the classical approach, a list of various counting techniques has been described. Some of the basic laws of probability such as Bayes' theorem and conditional probability have been provided. Although the readers are expected to have prior knowledge of probability before reading this book, this chapter may serve as a ready reference to the theory of probability to the prospective readers.

REFERENCES

1. Leung, K. T. and Chen, D. L.-C. (1967). *Elementary Set Theory: Parts I and II*. Hong Kong: Hong Kong University Press.
2. Rosen, K. H. (1998). *Discrete Mathematics and Applications*. Boston, MA: McGraw-Hill.

4 Probability Distributions

4.1 INTRODUCTION

The study of probability distributions is important for the study of simulation and the related software. As we all know, simulation study basically imitates the existing practical scenarios for the problems in various fields of study. Now, all the practical problems in real life involve uncertainties. These uncertainties may lie in the various aspects of a problem. The uncertainty lying within a problem can be handled in many ways. The existing literature shows at least three basic methods of dealing with the problems of uncertainty: application of probability theory, application of possibility theory, and application of fuzzy logic. Among these three methods, the theory of probability is basically used in all simulation software. This chapter discusses the various aspects and applications of various probability distributions.

The probability distributions in simulation software are used for various purposes. For example, the input parameters may follow certain kinds of probability distributions. The processing times, the waiting times in queue, the waiting times in buffers, the service times, the transportation times, the interarrival times, and the delays in processing may all follow separate probability distributions.

For example, let us consider a simple manufacturing process model. A part enters the production floor and waits to be processed by the first machine, and then after waiting for some time, the part is loaded onto the machine. After processing in the first machine, the part is unloaded and is again placed in a buffer to be processed on the next machine. Each machine takes some time to load the part, unload the part, and process the part, and there are also other times related to the machine, such as the setup time of the machine. Therefore, for this example, the probability distribution can be applied for the arrival times of the parts, the interarrival times of the parts, the waiting times in the queue, the waiting times in the buffers, the processing times on the machines, the loading times, the unloading times, the setup times, the transportation times, and so on.

Consider a bank teller as a second example. A customer arrives at the bank, chooses a particular queue in front of a particular clerk, and waits to receive the service. After receiving the service by the clerk, the customer leaves the counter. Here, the probability distribution can be used in several points in time, such as the arrival times of the customers, the interarrival times, the waiting times in the queue, and the service times of the clerk. The probability distribution is applied to these components since the times taken by all these components are uncertain. However, before going into the depths of probability distributions, let us explore first the concept of random variables, since we will be discussing the probability distributions of various random variables. Thus, Section 4.2 discusses the definition and various aspects of random variables.

The remaining sections of this chapter are arranged as follows: Section 4.2 introduces the concept of random variables; Section 4.3 describes the concepts of discrete and continuous probability distributions; Section 4.4 demonstrates the various discrete probability distributions; Section 4.5 depicts the various continuous probability distributions; and Section 4.6 provides the conclusion of this chapter.

4.2 INTRODUCTION TO RANDOM VARIABLES

The basic purpose of defining random variables is to use the concept of random variables to define the various probability functions. In most of the practical situations, we may not be interested in the entire physical description of the sample space or the events. Rather than that, we may be interested in certain numerical characteristics of the events. Suppose we have 10 instruments and they are operating for a certain amount of time. After working for a certain amount of time, we might like to know how many of them are working properly and how many of them are not working properly. If there are 10 instruments, it may happen that 8 of them are working properly and 2 of them are not working properly. At this stage, we may not be interested to know the positions and to know which of the two instruments have failed to work. Rather than that, we may be more interested to know the total number of instruments that have failed.

To effectively study the random phenomena or experiments, it is required to quantify the phenomena. This means that for every outcome, we associate a real number, and then, we can study the probability distribution of that. This brings us to the concept of random variables. Roughly speaking, if we give a notation x, then we let x be a function from Ω to R, that is, $x : \Omega \to R$. This means that for every outcome, we associate a real number, for example, runs scored by a cricket player, the yield of a particular crop, and the amount of rainfall in certain region.

Let (Ω, β, P) be a probability space. The function x is called a random variable if it is measurable. To explain the condition for measurability, suppose C is a class of subset of R so that C is a σ-field. For any $B \in \beta$, $x^{-1}(B) \in \beta$ and x defined on (Ω, β) is a random variable if and only if any one of the following conditions is satisfied:

- $\{\omega : x(\omega) \leq \lambda\} \in \beta, \forall \lambda \in R$
- $\{\omega : x(\omega) < \lambda\} \in \beta, \forall \lambda \in R$
- $\{\omega : x(\omega) \geq \lambda\} \in \beta, \forall \lambda \in R$
- $\{\omega : x(\omega) > \lambda\} \in \beta, \forall \lambda \in R$

From the above conditions, we can say that if x is a random variable, then we have the following as the consequences:

- $\{\omega : x(\omega) = \lambda\} \in \beta, \forall \lambda \in R$
- $\{\omega : \lambda_1 < x(\omega) \leq \lambda_2\} \in \beta, \forall \lambda_1 < \lambda_2$

Let us consider two simple examples based on the above concept.

Example 1

Let us consider the example of throwing two dice. Suppose x represents the sum of the numbers of the two dice. Every die has six faces numbered 1, 2, 3, 4, 5, and 6. However, the sample space for the two dice will be

$$\Omega = \{(1,1),(1,2),(1,3),(1,4),(1,5),(1,6),(2,1),(2,2),(2,3),$$

$$(2,4),(2,5),(2,6),(3,1),(3,2),(3,3),(3,4),(3,5),(3,6),$$

$$(4,1),(4,2),(4,3),(4,4),(4,5),(4,6),(5,1),(5,2),(5,3),$$

$$(5,4),(5,5),(5,6),(6,1),(6,2),(6,3),(6,4),(6,5),(6,6)\}$$

(4.1)

Let β = Power set(Ω), that is, β is the power set of Ω. Based on these data, we can have the following:

$$\{\omega : x(\omega) \leq \lambda\} = \begin{cases} \Phi, & \lambda \leq 1 \\ \{(1,1)\}, & \lambda \leq 2 \\ \{(1,1),(1,2),(2,1)\}, & \lambda \leq 3 \\ \{(1,1),(1,2),(2,1),(2,2),(1,3),(3,1)\}, & \lambda \leq 4 \\ \cdots & \cdots \\ \Omega, & \lambda \leq 12 \end{cases}$$

(4.2)

Thus, x is a random variable.

Example 2

Let us consider that a fair coin has been tossed thrice. Suppose the random variable x represents the number of heads.

Now, if the coin is tossed thrice, then the sample space can be represented by

$$\Omega = \{TTT, TTH, THT, THH, HTT, HTH, HHT, HHH\}$$ (4.3)

Based on these data, we can have the following:

$$\{\omega : x(\omega) \leq \lambda\} = \begin{cases} \{\Phi, TTT\}, & \lambda \leq 0 \\ \{TTT, TTH, THT, HTT\}, & \lambda \leq 1 \\ \{TTT, TTH, THT, THH, HTT, HTH, HHT\}, & \lambda \leq 2 \\ \Omega, & \lambda \leq 3 \end{cases}$$

(4.4)

The distribution of the random variable x can be defined through the cumulative distribution function (CDF), $F_{x'}(x) = P(x \leq x')$ for $-\infty < x < +\infty$. The behavior of the random variable x is determined by the CDF $F_{x'}(x)$. Thus, a random variable can be a discrete or continuous random variable.

A discrete random variable x defined on a probability space (Ω,β,P) is said to be discrete if there exists a countable set $E \subset R \rightarrow P(x \in E) = 1$, that is, the range of x is at most countable. The points of E that have positive mass are called jump points and the size of the jump is the probability of the random variable taking that point.

A random variable x is continuous if its CDF $F_{x'}(x)$ is absolutely a continuous function, that is, there exists a nonnegative function $f_{x'}(x)$ such that

$$F_{x'}(x) = \int_{-\infty}^{x} f_{x'}(t)dt, \ \forall x' \in R \tag{4.5}$$

The function $f_{x'}(x)$ is called the probability density function (PDF) of x. If F is absolutely continuous and f is continuous at x, then

$$\frac{d}{dx}F_{x'}(x) = f_{x'}(x) \tag{4.6}$$

The PDF satisfies the following:

$$f_{x'}(x) \geq 0, \ \forall x \in R \tag{4.7}$$

$$\int_{-\infty}^{\infty} f_{x'}(x)dx = 1 \tag{4.8}$$

$$\int_{a}^{b} f_{x'}(x)dx = P(a \leq x \leq b) \tag{4.9}$$

Based on the type of random variable (discrete or continuous), the probability distribution can also be a discrete or continuous probability distribution. Thus, Section 4.3 discusses discrete and continuous probability distributions.

4.3 DISCRETE AND CONTINUOUS PROBABILITY DISTRIBUTIONS

Random variables are so important in random experiments that we often do not emphasize them in the associated sample space at all and rather keep our focus on the distribution of the random variables. The probability distribution of a random variable x can be defined as the description of the probability values that are associated with the values of the random variable x [1]. The values of a discrete random variable can be distinguished from one another, that is, they are distinct values. The associated probability values will also be discrete in nature. Thus, a discrete probability distribution is associated with discrete variables. For example, "the probability that $x = 0$ is 0.34" can be expressed as $P(x = 0) = 0.34$. Similarly, the expression $P(x = 2) = 0.89$ means that the probability that $x = 2$ is 0.89. Thus, in these examples, the values of the random variable x are discrete (0 and 2), and naturally, each of these values is associated with a distinct probability value. Thus, the respective probability distribution is a discrete probability distribution.

On the contrary, a continuous random variable takes continuous values, for example, the weight of a particular product. Suppose the perfect weight of the product is 10 g. But the actual weight may vary to some extent, for various factors, such as wind speed and moisture content. In most cases, the weight will not be exactly 10 g. Instead, it may be 10.1 or 9.9 g or alike. Consider another example of the length of a metal bar. The length of the bar can vary for various reasons, such as temperature fluctuation, calibration errors, operator fault, and wearing. Thus, the length in these cases will vary within a certain range of values, rather than a single fixed value. In all these cases, the length or the weight is better represented by a range of values called an interval. However, since the range of possible values of the random variable in these cases is uncountably infinite, such a random variable is associated with a different kind of distribution. Such a distribution is continuous in nature because within a particular interval the random variable can have any value. For example, in the closed interval [2,3], we can assume any value if the respective variable contains real values. Thus, the values, for example, can be 2, 2.1, 2.4, 2.8, 2.533, 2.3436,... Therefore, a continuous random variable generally contains real numbers.

The description of the probability values for a discrete random variable is provided by a function that is known as the probability mass function (pmf). Thus, the pmf of a discrete random variable x with possible values $x_1, x_2, x_3, ..., x_n$ is given by the function $f(x)$, where

$$f(x_i) = P(X = x_i) \tag{4.10}$$

The pmf has the following properties:

- $f(x_i) \geq 0$. This is the basic condition of all the probability values. It says that the probability values should be positive.
- $\sum_{i=1}^{n} f(x_i) = 1$, that is, the sum of all the probability values is 1.

Example 3

Let the probabilities that a random variable X will have values 1, 2, 3, and 4 be 0.3, 0.1, 0.2, and 0.4, respectively, which can be represented by the following expressions:

$$f(1) = P(X = 1) = 0.3$$

$$f(2) = P(X = 2) = 0.1$$

$$f(3) = P(X = 3) = 0.2 \tag{4.11}$$

$$f(4) = P(X = 4) = 0.4$$

Thus, all of the above $f(x_i)$ expressions are positive quantity, that is, $f(x_i) \geq 0$. The sum of all $f(x_i)$ is $0.3 + 0.1 + 0.2 + 0.4 = 1.0$, thus satisfying the second property as stated earlier.

Another function applied in case of discrete probability distributions is called the CDF. In very simple terms, the CDF calculates the sum of a set of probability values that satisfy a particular condition specified in the definition of a CDF. Thus, the CDF of a discrete random variable X can be defined as

$$F(x) = P(X \leq x) = \sum_{x_i \in x} f(x_i) \tag{4.12}$$

The properties of a CDF for the discrete random variable X are provided as follows:

- The value of $F(x)$ will lie between 0 and 1, that is, $0 \leq F(x) \leq 1$.
- If $x \leq y$, then $F(x) \leq F(y)$.

We provide the following examples, which will clarify the above-mentioned concept of CDFs.

Example 4

Consider that a coin has been tossed thrice. Thus, the sample space can be given by

$$\Omega = \{TTT, TTH, THT, THH, HTT, HTH, HHT, HHH\} \tag{4.13}$$

Suppose that the random variable x denotes the event of getting a number of heads. Then, the probability of getting no head can be given by

$$f(x_0) = P(x = 0) = \frac{1}{8} \tag{4.14}$$

The probability of getting one head can be given by

$$f(x_1) = P(x = 1) = \frac{3}{8} \tag{4.15}$$

The probability of getting two heads can be given by

$$f(x_2) = P(x = 2) = \frac{3}{8} \tag{4.16}$$

The probability of getting three heads can be given by

$$f(x_3) = P(x = 3) = \frac{1}{8} \tag{4.17}$$

Thus, the total probability is given by

$$\sum_i f(x_i) = f(x_0) + f(x_1) + f(x_2) + f(x_3) = \frac{1}{8} + \frac{3}{8} + \frac{3}{8} + \frac{1}{8} = 1 \tag{4.18}$$

Now, we want to find out the CDFs. Thus, the CDF representing the number of occurrences of getting at most no head is given by

$$F(0) = P(x \leq 0) = f(x_0) = \frac{1}{8} \tag{4.19}$$

The CDF representing the number of occurrences of getting at most one head is given by

$$F(1) = P(x \leq 1) = f(x_0) + f(x_1) = \frac{1}{8} + \frac{3}{8} = \frac{1}{2} \tag{4.20}$$

The CDF representing the number of occurrences of getting at most two heads is given by

$$F(2) = P(x \leq 2) = f(x_0) + f(x_1) + f(x_2) = \frac{1}{8} + \frac{3}{8} + \frac{3}{8} = \frac{7}{8} \qquad (4.21)$$

The CDF representing the number of occurrences of getting at most three heads is given by

$$F(3) = P(x \leq 3) = f(x_0) + f(x_1) + f(x_2) + f(x_3) = \frac{1}{8} + \frac{3}{8} + \frac{3}{8} + \frac{1}{8} = 1 \qquad (4.22)$$

Thus, from the above calculation, we observe that all $0 \leq F(x) \leq 1$, which is the first property of a CDF. We can also find out the pmf if the CDF is given. The following example illustrates this fact.

Example 5

Let us consider the following CDF:

$$F(x) = \begin{cases} 0, & x < -2 \\ 0.2, & -2 \leq x < 0 \\ 0.7, & 0 \leq x < 2 \\ 1, & 2 \leq x \end{cases} \qquad (4.23)$$

We can calculate the pmf from the given CDF in the following way. Since CDF is the sum of the probability values up to a certain level, we can obtain the pmf by simply subtracting the later values of $F(x)$ from the previous values of $F(x)$. Therefore, we get the following values:

$$f(-2) = 0.2 - 0 = 0.2$$

$$f(0) = 0.7 - 0.2 = 0.5 \qquad (4.24)$$

$$f(2) = 1 - 0.7 = 0.3$$

These are the values for the respective pmf. Up to now, we have discussed the discrete probability distribution. The continuous probability distribution is also very interesting to delve into. Similar to the pmf, the continuous probability distribution is associated with a PDF, since the continuous distribution does not deal with individual discrete values.

The density function is widely used in physics. To measure the density of a material, we divide the mass of the material by its volume. This is almost equivalent to saying that the loading of the material is equally divided over the entire mass of the material when we divide its mass by its volume. Similarly, we calculate the PDF f(x) to describe the probability distribution over a continuous random variable x. This is equivalent to saying that the probability density is distributed over the entire interval of a continuous random variable.

For a continuous random variable X, the PDF can be defined as

$$P(a \leq X \leq b) = \int_a^b f(x)dx \qquad (4.25)$$

The above expression indicates the area defined by f(x) between the values a and b, for any values of a and b. The properties of a PDF are as follows:

- $f(x) \geq 0$, that is, the probability values are always positive.
- $\int_{-\infty}^{\infty} f(x)dx = 1$, that is, the sum of the total probability values is 1.

A PDF is represented by an area under a curve represented by f(x) and it is approximated by a histogram. Whenever we want to get a certain value of a PDF for a particular range of values, we actually find out the subset of the area represented by f(x) within the pair of given values. To understand the concept, let us consider the following two examples:

Example 6

Consider the PDF given by $f(x) = 5e^{-5(x-12)}$. We can find out the probability for $x > 12.8$.

The required probability can be found out in the following way:

$$P(X > 12.8) = \int_{12.8}^{\infty} f(x)dx = \int_{12.8}^{\infty} 5e^{-5(x-12)}dx = -e^{-5(x-12)} \, |_{12.8}^{\infty} = 0.0183 \quad (4.26)$$

Example 7

Consider the PDF given by $f(x) = 0.4$ for $0 \leq x < 10$.

To find out, for example, the probability that $x < 6$, that is, $P(x < 6)$, we will have to perform the following simple calculation:

$$P(X < 6) = \int_0^6 f(x)dx = \int_0^6 0.4 dx = 2.4 \qquad (4.27)$$

The CDF for the continuous random variable X is given by

$$F(x) = P(X \leq x) = \int_{-\infty}^{x} f(t)dt \text{ for } -\infty < x < \infty \qquad (4.28)$$

An example for the CDF is provided as follows:

Example 8

Similar to Example 7, we can calculate the CDF.
Thus, $F(x)$ will be

$$F(x) = \int_0^x f(t)dt = \int_0^x 0.4dt = 0.4x, \text{ for } 0 \leq x < 10 \qquad (4.29)$$

For $x \geq 10$, we have

$$F(x) = \int_0^x f(t)dt = 1 \qquad (4.30)$$

Thus, the CDF is given by

$$F(x) = \begin{cases} 0, & x < 0 \\ 0.4x, & 0 \leq x < 10 \\ 1, & 10 \leq x \end{cases} \qquad (4.31)$$

The above given concepts are the basic concepts on which all the various probability distributions are based. Thus, Section 4.4 first describes some of the discrete probability distributions.

4.4   VARIOUS DISCRETE PROBABILITY DISTRIBUTIONS

Before discussing the various discrete probability distributions, let us have a look at some of the numerical properties of discrete random variables. The properties are given as follows:

- The mean of the expectation of a discrete random variable X is denoted by μ and is given by

$$\mu = E(X) = \sum_x xf(x) \qquad (4.32)$$

- The variance of a discrete random variable X is denoted by σ^2 or $\text{Var}(X)$ and is given by

$$\sigma^2 = \text{Var}(X) = E(X-\mu)^2 = \sum_x (X-\mu)^2 f(x) \qquad (4.33)$$

- The standard deviation of a discrete random variable X is given by

$$\sigma = \sqrt{\sigma^2} \qquad (4.34)$$

Let us consider the following example to understand the above concepts:

Example 9

Consider Example 3. The probability values are given as follows:

$$f(1) = P(X = 1) = 0.3$$

$$f(2) = P(X = 2) = 0.1$$

$$f(3) = P(X = 3) = 0.2 \qquad (4.35)$$

$$f(4) = P(X = 4) = 0.4$$

Thus, the mean of these values can be found as

$$\mu = E(X) = \sum_x xf(x) = (1 \times 0.3) + (2 \times 0.1) + (3 \times 0.2) + (4 \times 0.4) = 2.7 \quad (4.36)$$

The variance can be calculated in the following way. The calculations are given in Table 4.1.

Thus, the variance is

$$\sigma^2 = \text{Var}(X) = \sum_x (X - 2.7)^2 f(x) \qquad (4.37)$$

$$= 0.867 + 0.049 + 0.018 + 0.676 = 1.61$$

Thus, the standard deviation will be

$$\sigma = 1.269 \qquad (4.38)$$

Now, we are ready to briefly present the various discrete probability distributions in Sections 4.4.1 through 4.4.6. The distributions that are presented here are as follows:

- Discrete uniform distribution
- Binomial distribution
- Geometric distribution
- Negative binomial distribution
- Hypergeometric distribution
- Poisson distribution

TABLE 4.1

Calculation of Variance for Example 9

x	$x-2.7$	$(x-2.7)^2$	$f(x)$	$f(x)\,(x-2.7)^2$
1	−1.7	2.89	0.3	0.867
2	−0.7	0.49	0.1	0.049
3	0.3	0.09	0.2	0.018
4	1.3	1.69	0.4	0.676

4.4.1   DISCRETE UNIFORM DISTRIBUTION

The discrete uniform distribution is the simplest of all discrete probability distributions. If a random variable X assumes its values $x_1, x_2, ..., x_n$ with equal probability, then the probability distribution is called a discrete uniform probability distribution.

Thus, the pmf of the discrete uniform distribution is given by

$$f(x_i) = \frac{1}{n} \qquad (4.39)$$

The mean of this distribution within the range $[a,b]$ is given by

$$\mu = E(X) = \frac{a+b}{2} \qquad (4.40)$$

The variance of this distribution is given by

$$\sigma^2 = \frac{(b-a+1)^2 - 1}{12} \qquad (4.41)$$

The mode is not applicable for this distribution. The discrete uniform probability distribution is shown in Figure 4.1.

The following numerical example can illustrate the above concept.

Example 10

Consider the tossing of a fair die. Then, the sample space is given by $S = \{1, 2, 3, 4, 5, 6\}$.

Thus, we will have discrete uniform distribution with pmf as

$$f(x_i) = \frac{1}{6} \qquad (4.42)$$

The mean of this distribution will be

$$\mu = E(X) = \sum_x x f(x)$$

$$\qquad (4.43)$$

$$= \left(1 \times \frac{1}{6}\right) + \left(2 \times \frac{1}{6}\right) + \left(3 \times \frac{1}{6}\right) + \left(4 \times \frac{1}{6}\right) + \left(5 \times \frac{1}{6}\right) + \left(6 \times \frac{1}{6}\right) = 3.5$$

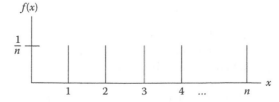

FIGURE 4.1   Discrete uniform probability distribution.

TABLE 4.2

Calculation of Variance for Example 10

x	x−3.5	(x−3.5)²	f(x)
1	−2.5	6.25	1/6
2	−1.5	2.25	1/6
3	−0.5	0.25	1/6
4	0.5	0.25	1/6
5	1.5	2.25	1/6
6	2.5	6.25	1/6

The calculations for variance are given in Table 4.2.

Thus, the variance is given by

$$\sum f(x)(x-3.5)^2 = \frac{1}{6}(6.25 + 2.25 + 0.25 + 0.25 + 2.25 + 6.25) \tag{4.44}$$

$$= 2.92$$

Discrete uniform distributions can be used in nonparametric statistics, where the samples do not depend on a parameter and the distribution of the sample is unknown. These types of statistics are more robust than parametric methods because they do not assume structure; they are nonbiased. The discrete uniform distribution can be applied when statistics are based on the ranks of events. For example, in a data set consisting of descriptive terms such as hot, cold, or warm, rank statistics would assign numerical values to each observation as 3, 1, and 2, respectively.

4.4.2 BINOMIAL DISTRIBUTION

Suppose there are n independent trials. A binomial distribution is the number of times a particular event occurs with probability p. Thus, p is the probability of an event in a single trial. Examples of such experiments may be as follows:

- Tossing a coin for n times
- Tossing n number of coins
- Number of x students passed in a group of n students
- Independent throwing of n dice

The pmf for the binomial distribution is given by

$$f(x) = {}^nC_x p^x q^{n-x} \tag{4.45}$$

where p is the probability of the event and $p + q = 1$, and thus, $q = 1 - p$. The general view of a binomial distribution is provided in Figure 4.2.

Some of the important properties of binomial distribution are given as follows:

- The random variables in a binomial distribution assume discrete values.
- The parameters of a binomial distribution are n and p.

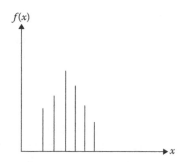

FIGURE 4.2   Binomial distribution.

- The mean of a binomial distribution is given by $\mu = E(X) = np$.
- The variance of a binomial distribution is given by $\sigma^2 = npq$.
- A binomial distribution may have one or two modes. When $(n+1) \times p$ is not an integer, the mode will be the largest integer contained. When it is an integer, there are two modes—$(n+1) \times p$ and $(n+1) \times p - 1$.
- If x and y follow a binomial distribution with parameters (n_1, p) and (n_2, p), respectively, then $(x + y)$ also follows a binomial distribution with parameters $(n_1 + n_2, p)$.
- A binomial distribution may be a limiting case for hypergeometric distribution.

A binomial distribution is applicable in only those cases where the following conditions are satisfied:

- The result of any trial can be categorized into two types only.
- The probability of success in each trial is fixed.
- The trials are independent.

The following example illustrates the application of binomial distribution.

Example 11

A fair coin is tossed 10 times. What is the probability of getting six heads? Assume that the probability of getting a head is 1/2.

In this example, the number of trials is $n = 10$. The probability of getting a head is $p = 1/2$, and thus the probability of not getting a head is $q = 1 - p = 1 - (1/2) = 1/2$. Thus, the required probability is $f(x) = {}^{10}C_6 (1/2)^6 (1/2)^{10-6} = 0.205$.

The mean for this problem is $\mu = E(X) = np = 10 \times (1/2) = 5$, and the variance is $\sigma^2 = npq = 10 \times (1/2) \times (1/2) = 2.5$.

The conditions for Bernoulli distributions are same as those for binomial distributions. The number X of successes in n Bernoulli trials is called a binomial random variable. For example, X could be the number of heads in n tosses of a coin. Due to similar conditions and expressions, we are not presenting any subsection on Bernoulli distribution.

4.4.3 Geometric Distribution

If X is a random variable of the number of trials up to and including the first success, then the distribution of X is called geometric distribution. The pmf for the geometric distribution is given by

$$f(x) = (1-p)^{n-1} p \qquad (4.46)$$

Thus, in the case of a geometric distributions, we keep on continuing the trials until we get a success. For example, if we keep on tossing a coin a number of times until a head appears, then the probability of getting the first head will be calculated by applying the geometric distribution. The characteristics of geometric distributions are as follows:

- The trials are n independent Bernoulli trials.
- The number of trials keeps on increasing until a success is obtained.
- The parameter of a geometric distribution is p, where p is the probability of success.
- The probability of success in each trial is fixed.
- The mean of geometric distribution is $\mu = E(X) = 1/p$.
- The variance of geometric distribution is $\sigma^2 = (1-p)/p^2$.
- A geometric distribution shows a memoryless property, that is, the success in a trial is independent of the outcomes of the previous trials.
- Among all the discrete probability distributions, the geometric distribution shows the largest entropy.

A geometric distribution function is shown in Figure 4.3.

The following numerical example describes the application of a geometric distribution.

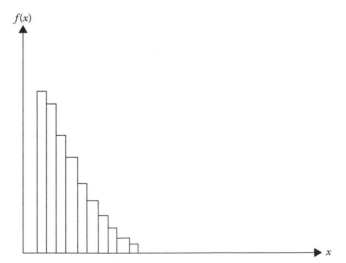

FIGURE 4.3 Geometric distribution.

Example 12

A coin is tossed several times until a head appears. What is the probability of getting a success after four trials? Assume that the probability of getting a head is 2/3.

Thus, the number of trials after which the head appears is four and the total number of trials is $n = 5$. The probability of getting a head is 2/3, that is, $p = 2/3$.

Thus, the required probability is

$$f(x) = (1-p)^{n-1}p = \left(1 - \frac{2}{3}\right)^{5-1}\left(\frac{2}{3}\right) = \frac{2}{243} \qquad (4.47)$$

4.4.4 Negative Binomial Distribution

A negative binomial distribution is a generalization of a geometric distribution. The random variable in a negative binomial distribution is the number of Bernoulli trials required to obtain r number of successes. The pmf of a negative binomial distribution is as follows:

$$f(x) = {}^{x-1}C_{r-1}(1-p)^{x-r}p^r \qquad (4.48)$$

The characteristics of negative binomial distributions are as follows:

- Independent Bernoulli trials are performed.
- The mean of a negative binomial distribution is given by $\mu = E(X) = r/p$.
- The variance of a negative binomial distribution is $\sigma^2 = r(1-p)/p^2$.
- A negative binomial distribution is the number of trials required to get r successes. Thus, the number of successes r is prespecified. Due to this fact, the negative binomial distribution can be considered as opposite or the negative of a binomial random variable.
- The geometric distribution is a special case of a negative binomial distribution.

The following example helps to build up the concept of negative binomial distributions.

Example 13

Let X denote the number of requests until all three servers fail, and let X_1, X_2, and X_3 denote the number of requests before the failure of the first, second, and third servers used, respectively. Now, $X = X_1 + X_2 + X_3$. Also, the requests are assumed to comprise independent trials with a constant probability of failure $p = 0.0005$. Furthermore, a spare server is not affected by the number of requests before it is activated. Therefore, X has a negative binomial distribution with $p = 0.0005$ and $r = 3$. Consequently, $E(X) = 3/0.0005 = 6000$ requests.

What is the probability that all three servers fail within five requests?

The respective probability is

$$P(X \leq 5) = P(X = 3) + P(X = 4) + P(X = 5)$$

$$= (0.0005)^3 + {}^3C_2(0.0005)^3(0.9995) + {}^4C_2(0.0005)^3(0.9995)^2$$

$$= 1.249 \times 10^{-9}$$

4.4.5 HYPERGEOMETRIC DISTRIBUTION

Suppose an urn contains N balls in which A balls are red and $N - A$ balls are yellow. If n balls are selected at random from the total N balls, then the probability of getting x red balls and $n - x$ yellow balls is given by

$$f(x) = \frac{{}^A C_x \, {}^{N-A}C_{n-x}}{{}^N C_n} \tag{4.49}$$

This equation is the pmf for a hypergeometric distribution. The properties of hypergeometric distributions are given as follows:

- The random variable in a hypergeometric distribution consists of a finite number of values, that is, $x = 0,1,2,...,m$.
- A hypergeometric distribution has three parameters—A, N, and n.
- The mean of a hypergeometric distribution is given by $\mu = E(X) = nA/N$.
- The variance of a hypergeometric distribution is given by

$$\sigma^2 = \frac{nA(N-A)(N-n)}{N^2(N-1)}$$

- The sample size of n objects is selected at random without replacement.
- A objects are classified as successes and $N - A$ objects are classified as failures.
- The value of x ranges between $\max\{0, n+A-N\}$ and $\min\{A, n\}$.
- A binomial distribution may be obtained as a limiting case of a hypergeometric distribution when $N \to \infty$.

A hypergeometric distribution is shown in Figure 4.4.

The following numerical example may clarify the idea more distinctly.

Example 14

A bag contains 10 red balls and 20 blue balls. Eight balls are drawn at random from the bag without replacement. Calculate the probability that five among them are red.

In this example, the total number of balls is $N = 10 + 20 = 30$. Among these, the number of red balls is $A = 10$ and the number of blue balls is $N - A = 20$.

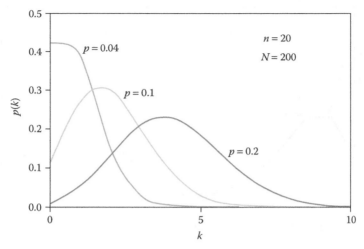

FIGURE 4.4 An example of a hypergeometric distribution.

A total of eight balls have been chosen at random, that is, $n = 8$. Among these balls, $x = 5$ balls will have to be red. Thus, the required probability is

$$f(x) = \frac{{}^A C_x \, {}^{N-A} C_{n-x}}{{}^N C_n} = \frac{{}^{10} C_5 \, {}^{20} C_3}{{}^{30} C_8} \tag{4.50}$$

4.4.6 POISSON DISTRIBUTION

The number of outcomes occurring in a specific time interval is given by a random variable for a Poisson distribution. The pmf for a Poisson distribution is

$$f(x) = \frac{e^{-m} m^x}{x!} \text{ for } x = 0, 1, 2, \dots, \infty \tag{4.51}$$

where:
  m is positive

The properties of Poisson distributions are as follows:

- The random variable in a Poisson distribution assumes countably infinite values.
- The mean of random variable for a Poisson distribution is $\mu = E(x) = m$.
- The variance of the random variable for a Poisson distribution is $\sigma^2 = \mathrm{Var}(x) = m$.
- A Poisson distribution may have either one or two modes. The value of the mode will be the largest integer contained in m if m is not an integer. Otherwise, if m is an integer, then the modes will be m and $m - 1$.
- If the Poisson random variables x and y have the parameters m_1 and m_2, then $(x + y)$ will also follow a Poisson distribution with the parameter $(m_1 + m_2)$.

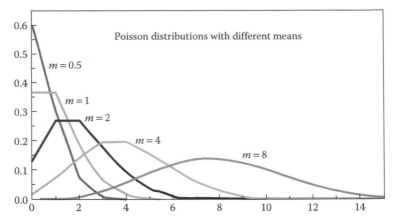

FIGURE 4.5 A graph for Poisson distribution.

- A Poisson distribution may be used as an approximation to a binomial distribution, when p is small and n is large.

The Poisson distribution is shown in Figure 4.5.

The following simple numerical example may clarify the calculations for Poisson distributions.

Example 15

Consider a certain random variable x following a Poisson distribution. The parameter for the distribution is 4. Find the probability that the variable will assume the value of 2.

In this example, the parameter value is $m = 4$. Thus, the required probability value will be

$$f(x) = \frac{e^{-m}m^x}{x!} = \frac{e^{-4}4^2}{2!} = 0.1465 \qquad (4.52)$$

Section 4.5 shows some continuous probability distributions.

4.5   VARIOUS CONTINUOUS PROBABILITY DISTRIBUTIONS

Among a variety of continuous probability distributions, this section discusses the following:

- Exponential distribution
- Beta distribution
- Erlang distribution

- Gamma distribution
- Johnson distribution
- Normal distribution
- Lognormal distribution
- Triangular distribution
- Weibull distribution

Before discussing the above continuous probability distributions, certain numerical properties of a continuous random variable are given as follows:

- The mean of a continuous random variable with PDF $f(x)$ is given by

$$\mu = E(X) = \int_{-\infty}^{\infty} xf(x)\mathrm{d}x \tag{4.53}$$

- The variance of a continuous random variable is

$$\sigma^2 = \mathrm{Var}(x) = \int_{-\infty}^{\infty} (x-\mu)^2 f(x)\mathrm{d}x = \int_{-\infty}^{\infty} x^2 f(x)\mathrm{d}x - \mu^2 \tag{4.54}$$

To explain the concept, let us consider the following example:

Example 16

Consider the PDF $f(x) = 0.4x$ for $0 \le x < 10$. The respective mean for this distribution will be

$$\mu = E(X) = \int_0^{10} xf(x)\mathrm{d}x = \int_0^{10} (x)0.4x\mathrm{d}x = \int_0^{10} 0.4x^2\mathrm{d}x = \frac{0.4}{3}x^3 \big|_0^{10} = 133.33 \tag{4.55}$$

The variance of the random variable is given by

$$\sigma^2 = \mathrm{Var}(x) = \int_0^{10} (x-\mu)^2 f(x)\mathrm{d}x = \int_0^{10} (x-133.33)^2 0.4x\mathrm{d}x = 320903.11 \tag{4.56}$$

Table 4.3 provides the PDF, CDF, mean, and variance of various continuous probability distributions mentioned earlier. Tables 4.4 and 4.5 show their corresponding appearances and application areas, respectively.

TABLE 4.3
PDF, CDF, Mean, and Variance

Probability Distribution	PDF	CDF	Mean	Variance
Exponential distribution	$f(x) = \begin{cases} \lambda e^{-\lambda x}, & x \geq 0 \\ 0, & \text{otherwise} \end{cases}$	$F(x) = \int_{-\infty}^{x} f(x)\,dx = \begin{cases} 1-e^{-\lambda x}, & x \geq 0 \\ 0, & \text{otherwise} \end{cases}$	$\dfrac{1}{\lambda}$	$\dfrac{1}{\lambda^2}$
Beta distribution	$f(x) = \begin{cases} \dfrac{x^{\alpha-1}(1-x)^{\beta-1}}{B(\alpha,\beta)}, & 0<x<1 \\ 0, & \text{otherwise} \end{cases}$, where B is the complete beta function and is given by $B(\alpha,\beta) = \int_0^1 t^{\beta-1}(1-t)^{\alpha-1}\,dt$	$F(X;\alpha,\beta) = \dfrac{B(X;\alpha,\beta)}{B(\alpha,\beta)} = I_x(\alpha,\beta)$, where $B(X;\alpha,\beta)$ is the complete beta function and $I_x(\alpha,\beta)$ is the regularized incomplete beta function	$\dfrac{\alpha}{\alpha+\beta}$	$\dfrac{\alpha\beta}{(\alpha+\beta)^2(\alpha+\beta+1)}$
Gamma distribution	$f(x) = \begin{cases} \dfrac{\beta^{-\alpha}x^{\alpha-1}e^{-x/\beta}}{\Gamma(\alpha)}, & x>0 \\ 0, & \text{otherwise} \end{cases}$, where Γ is the complete gamma function given by $\Gamma(\alpha) = \int_0^\infty t^{\alpha-1}e^{-t}\,dt$	$\dfrac{\gamma\left(\alpha,\dfrac{x}{\beta}\right)}{\Gamma(\alpha)}$, where $\gamma\left(\alpha,\dfrac{x}{\beta}\right)$ is the lower incomplete gamma function	$\alpha\beta$	$\alpha\beta^2$
Normal distribution	$f(x) = \dfrac{1}{\sigma\sqrt{2\pi}}\,e^{-\frac{(x-\mu)^2}{2\sigma^2}}$, where μ and σ are the mean and standard deviation, respectively. For $\mu = 0$ and variance σ^2, the distribution is called standard normal distribution.	$\Phi(x) = \dfrac{1}{\sqrt{2\pi}}\int_{-\infty}^{x} e^{-\frac{t^2}{2}}\,dt$ (for standard normal distribution)	μ	σ^2

Distribution	$f(x)$	$F(x)$	Mean	Variance
Lognormal distribution	$f(x) = \begin{cases} \dfrac{1}{\sigma x\sqrt{2\pi}}\,e^{-[\ln(x)-\mu]^2/(2\sigma^2)}, & x > 0 \\ 0, & \text{otherwise} \end{cases}$	$F(X) = \dfrac{1}{2}\left[1 + \text{erf}\left(\dfrac{\ln x - \mu}{\sigma\sqrt{2}}\right)\right]$ $= \dfrac{1}{2}\text{erfc}\left(-\dfrac{\ln x - \mu}{\sigma\sqrt{2}}\right) = \Phi\left(\dfrac{\ln x - \mu}{\sigma}\right),$ where erf is the error function and erfc is the complementary error function	$e^{\mu+\frac{1}{2}\sigma^2}$	$(e^{\sigma^2}-1)[E(x)]^2$
Triangular distribution	$f(x) = \begin{cases} 0, & x < a \\ \dfrac{2(x-a)}{(b-a)(c-a)}, & a \le x \le c \\ \dfrac{2(b-x)}{(b-a)(b-c)}, & c < x \le b \\ 0, & b < x \end{cases}$ where a, b, and c are constants; $a < b$ and $c \le b$	$F(x) = \begin{cases} 0, & x < a \\ \dfrac{(x-a)^2}{(b-a)(c-a)}, & a \le x \le c \\ 1 - \dfrac{(b-x)^2}{(b-a)(b-c)}, & c < x \le b \\ 1, & b < x \end{cases}$	$\dfrac{a+b+c}{3}$	$\dfrac{a^2+b^2+c^2+ab+bc+ca}{18}$
Weibull distribution	$f(x) = \begin{cases} \alpha\beta^{-\alpha}x^{\alpha-1}e^{-(x/\beta)^\alpha}, & x > 0 \\ 0, & \text{otherwise} \end{cases}$	$F(X) = \begin{cases} 1-e^{-(x/\beta)^\alpha}, & x \ge 0 \\ 0, & \text{otherwise} \end{cases}$	$\dfrac{\beta}{\alpha}\Gamma\left(\dfrac{1}{\alpha}\right)$	$\dfrac{\beta^2}{\alpha}\left[2\Gamma\left(\dfrac{2}{\alpha}\right) - \dfrac{1}{\alpha}\left[\Gamma\left(\dfrac{1}{\alpha}\right)\right]^2\right]$
Erlang distribution	$f(x) = \begin{cases} \dfrac{\beta^k x^{k-1} e^{-x/\beta}}{(k-1)!}, & x > 0 \\ 0, & \text{otherwise} \end{cases}$	$1 - \displaystyle\sum_{n=0}^{k-1} \dfrac{1}{n!}e^{-\beta x}(\beta x)^n$	$k\beta$	$k\beta^2$

TABLE 4.4

Appearances of Various Continuous Probability Distributions

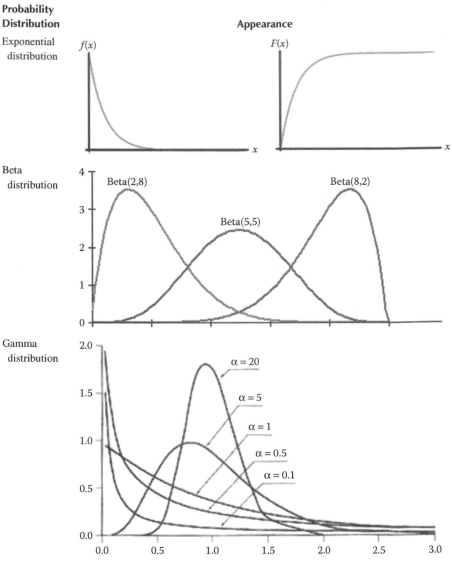

(*Continued*)

TABLE 4.4

(Continued) Appearances of Various Continuous Probability Distributions

Probability Distribution	Appearance
Normal distribution	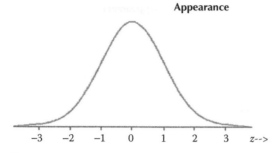
Lognormal distribution	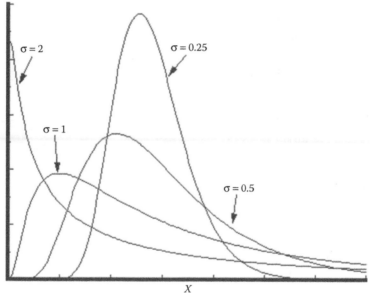
Triangular distribution	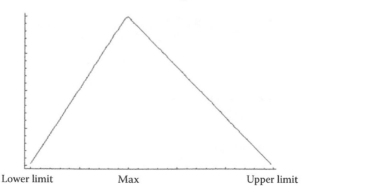

TABLE 4.4

(Continued) Appearances of Various Continuous Probability Distributions

Probability Distribution	Appearance

Weibull distribution

Erlang distribution

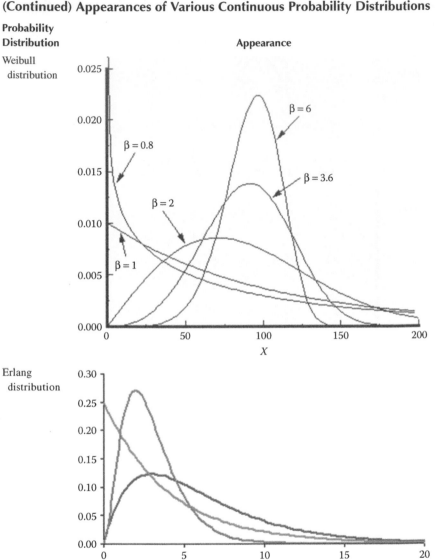

TABLE 4.5
Applications of Various Continuous Probability Distributions

Probability Distribution	Applications
Exponential distribution	This distribution is often used to model intervene times in random arrival and breakdown processes, but it is generally inappropriate for modeling process delay times.
Beta distribution	Due to its ability to take on a wide variety of shapes, this distribution is often used as a rough model in the absence of data. Because the range of the beta distribution is from 0 to 1, the sample X can be transformed to the scaled beta sample Y with the range from a to b by using the equation $Y = a + (b - a)X$. The beta is often used to represent random proportions, such as the proportion of defective items in a lot.
Gamma distribution	For integer shape parameters, this distribution is the same as the Erlang distribution. The gamma is often used to represent the time required to complete some task (e.g., machining time or machine repair time).
Normal distribution	This distribution is used in situations in which the central limit theorem applies, that is, quantities that are the sums of other quantities. It is also used empirically for many processes that appear to have a symmetric distribution.
Lognormal distribution	This distribution is used in situations in which the quantity is the product of a large number of random quantities. It is also frequently used to represent task times that have a distribution skewed to the right. This distribution is related to the normal distribution.
Triangular distribution	This distribution is commonly used in situations in which the exact form of the distribution is not known, but estimates for the minimum, maximum, and most likely values are available. This distribution is easy to use and explain than other distributions that may be used in this situation.
Weibull distribution	This distribution is widely used in reliability models to represent the lifetime of a device. If a system consists of a large number of parts that fail independently, and if the system fails when any single part fails, then the time between successive failures can be approximated by a Weibull distribution. This distribution is also used to represent nonnegative task times that are skewed to the left.
Erlang distribution	This distribution is used in situations in which an activity occurs in successive phases and each phase has an exponential distribution. For large k, this distribution approaches a normal distribution. This distribution is often used to represent the time required to complete a task and is a special case of the gamma distribution in which the shape parameter α is an integer (k).

4.6 CONCLUSION

In this chapter, a brief overview of the probability distribution concepts has been discussed. Separate sections have been devoted to both discrete and continuous probability distributions. The distributions have been described graphically and with numerical examples, especially for the discrete probability distributions. A number of discrete and continuous probability distributions have been discussed in detail. The readers are expected to get a brief reference to the common probability distribution used in various simulation software.

REFERENCE

1. Gun, A. M., Gupta, M. K., and Dasgupta, B. (2008). *Fundamentals of Statistics, Volume I.* Kolkata, India: World Press.

5 Introduction to Random Number Generators

5.1 INTRODUCTION

The basics of any stochastic simulation consist of generating a sequence of numbers that are independent and identically distributed with the common uniform distribution in [0,1]. Therefore, the basic ingredient of any stochastic simulation system is the generation of random numbers. Although in nature we have several sources of generating random numbers, to do any systematic analysis of a system, the generation of a stream of random numbers is required. These random numbers are basically generated using computer application software. Since any software package follows predefined algorithms, the computer-generated random numbers are not exactly random in nature. However, the numbers are generated in such a way that those numbers imitate the behaviors of true random numbers.

This chapter presents various methods of generating random numbers. Section 5.2 discusses the characteristics of a good random number generator and Section 5.3 discusses the different types of random numbers.

5.2 CHARACTERISTICS OF A RANDOM NUMBER GENERATOR

Random number generators are used for all sorts of computer simulation systems. Some of the real-life applications of random number generators include experimentations on traffic analysis, network systems, weather conditions, climate, Monte Carlo simulations, social studies, and so on. Therefore, knowledge of the characteristics of random number generators is required for the effective generation and use of random numbers.

The characteristics of a good random number generator are as follows:

1. *Independence.* First of all, a random number generator should be capable of generating numbers that are independent of each other and thus must not have any correlation among them.
2. *Uniform distribution.* The numbers generated should be distributed uniformly over [0,1].
3. *Long sequence.* The length of the sequence of a stream of random numbers generated should be very high. This means that a particular stream of random numbers should not recur over a long period of time.
4. *Repeatability.* The generator should provide some means so that a stream of random numbers can be regenerated. This sometimes helps in reproducing

an experiment or for repeated verification purposes or better comparison among simulations of different systems.

5. *Speed and resource.* The generator must be fast and must utilize as little storage as possible. This is possible when generating random numbers using computer software.

6. *Variety.* The generator must be capable of producing varieties of different streams of random numbers.

The above characteristics are the ideal ones. Thus, all random number generators may not have all the characteristics. However, the random numbers are usually generated in such a way that satisfies the above characteristics as much as possible. The characteristics of the random numbers and the generators depend on the type of random number generator used. Section 5.3 introduces the various classifications of random number generators.

5.3 TYPES OF RANDOM NUMBER GENERATORS

There is no demonstrated classification of random number generators. However, based on the method or means of generating random numbers as well as the characteristics of the generated numbers, random number generators can be classified into the following types:

- True random number generators
- Pseudorandom number generators
- Quasi-random number generators

While classifying random number generators, the main characteristic considered is "independence." Therefore, for true random number generators, the generated random numbers should exhibit true randomness. This means that the generated numbers do not follow any particularly defined algorithm, and thus these numbers are completely unpredictable. It is difficult to find correlation among the random numbers generated by true random number generators. Therefore, these random numbers cannot be regenerated.

The random numbers produced by pseudorandom number generators appear to be random but have a specific repetitive pattern. Since each pseudorandom number generator follows a specific algorithm that generates the random numbers, these random numbers can be regenerated. Quasi-random number generators generate a set of nonrandom numbers in a randomized fashion. A question about the utility of such random number generators arises automatically.

To understand all the above-mentioned random number generators, a detailed discussion on each type of random number generator is required. Section 5.3.1 first discusses true random number generators.

5.3.1 TRUE RANDOM NUMBER GENERATORS

True random number generators generate random numbers that show true randomness in terms of independence and unpredictability. These are basically random number machines or physical devices that generate random numbers based on various physical phenomena. Such physical phenomena may include thermal noise, photoelectric effect, or quantum phenomenon. The random numbers are generated from a low-level statistically random noise signal. The outputs of these processes are completely unpredictable and thus nonrepetitive.

These devices generally consist of a converter (such as a transducer) to convert other types of signals into electrical signals. An amplifier may be used to amplify or increase the signal. And if the output signal is analog in nature, then an analog-to-digital converter is required to get the output consisting of [0,1]. If properly sampled, such outputs generate streams of random numbers. The physical devices used are plugged into a computer that uses software to accept and use the generated random numbers.

A variety of physical phenomena have been used in the past for generating true random numbers. Some of these phenomena are as follows:

1. Various quantum mechanical concepts have been used for generating random numbers. Some of the quantum phenomena used for the purpose are given as follows:
   a. Nuclear decay radiation that is completely unpredictable
   b. "Shot noise" that is a quantum mechanical noise in electronic circuits
   c. Photons traveling through a semitransparent mirror
2. Counting of gamma rays
3. Biometric random number generators based on the phenomena such as the following:
   a. Animal neurophysiological brain responses
   b. Human galvanic skin responses [1]
4. Random number generators based on incoherent light by a fiber amplifier [2]
5. Thermal noise from a resistor
6. Atmospheric noise
7. Avalanche noise
8. Rapidly spinning disk
9. Pulsating of a vacuum tube that is a random phenomenon
10. Roulette wheel

Since such signals are difficult to catch and use, special physical devices have been built to generate and use the generated random numbers. Some examples of these devices are mentioned as follows:

1. Quantis card generating random numbers based on a quantum optics process
2. Electronic random number indicator equipment (ERNIE)
3. Geiger counter generating random numbers based on radioactive decay
4. Optical parametric oscillator connected to a PC

However, according to Tsoi et al. [3], there are three types of hardware implementations of physical random number generators, that is, oscillator sampling, direct amplification, and discrete-time chaos. Oscillator sampling is based on the period variation of a low-frequency clock; direct amplification is based on thermal or shot noise; and discrete-time chaos is based on the behavior of chaotic systems.

However, true random number generators are particularly preferable over pseudorandom number generators, especially for cryptography, gambling machines, and lotteries. These generators may be used for generating a seed value for a pseudorandom number generator, since the pseudorandom numbers are easy to track if the respective algorithm and the seed value can be found out somehow.

Most natural phenomena are unpredictable in nature and therefore can serve as effective sources for random numbers. However, the most difficult problem for random number generation is to find a way to use these random numbers in computers. Thus, the disadvantages of the true/physical random number generators are as follows:

1. True random number generators are very slow in generating random numbers.
2. The random numbers generated by true random number generators are not repeatable and cannot be regenerated.
3. Most devices are difficult to connect to PCs
4. The generated random numbers do not have a seed value, and thus each simulation is different from the other simulations.
5. Any significant occurrence is not reproducible.
6. Most physical devices produce numbers with a systematic bias.

Because of the above disadvantages, true random number generators are not used frequently. The most frequently used random number generators are pseudorandom number generators. Section 5.3.2 discusses the various pseudorandom number generators.

5.3.2 PSEUDORANDOM NUMBER GENERATORS

The word "pseudo" means that something is not the thing it is claimed to be. This means that pseudorandom numbers are not actual random numbers. In fact, pseudorandom numbers are generated by some formula or algorithm. Since the methods of generating the numbers are known, the true randomness of the numbers is vanished. However, the generated numbers can be regenerated if the method and the initial value (seed) are known. The numbers are generated in such a way that they appear to be random. The characteristics of pseudorandom numbers can be delineated through the following points:

1. The numbers generated by pseudorandom number generators appear to be random.
2. A stream of random numbers generated by pseudorandom number generators can be regenerated if the method and the initial value (seed) are known.
3. The generated numbers may not be distributed uniformly over [0,1].
4. There may be correlations among the numbers generated by pseudorandom number generators.

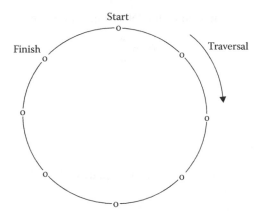

FIGURE 5.1 Random number cycle.

5. Since the numbers are not uniformly distributed, the mean and variance of the generated numbers may be either very high or very low.
6. The numbers are generated in cycles. This means that, for a given seed value, the same set of numbers is repeated (Figure 5.1). Thus, the sequence of random numbers has a finite number of integers.

Therefore, while developing a random number generator, some issues must be taken care of. These issues are as follows:

1. The sequence (or cycle) of random numbers should be long.
2. The generated numbers should be uniformly distributed over [0,1] so that the mean and variance are neither very high nor very low.
3. The algorithm should be fast in generating the random numbers.

There are numerous random number generators. Some of the most prominent ones are as follows:

1. Linear congruential generator (LCG)
2. Multiplicative congruential generator
3. Inversive congruential generator
4. Combined LCG
5. Lagged Fibonacci generator
6. Mid-square method

These methods are described in Sections 5.3.2.1 through 5.3.2.6.

5.3.2.1 Linear Congruential Generator

The expression for an LCG is as follows [1]:

$$x_{n+1} = (ax_n + c) \bmod m \tag{5.1}$$

The random number is generated by the following expression:

$$R_n = \frac{x_n}{m} \tag{5.2}$$

where:

    a, c, and m are constants
    a is the multiplier
    c is the increment
    m is the modulus

The values of these constants determine the quality of the LCG. The initial value of x_n (x_0) is called the "seed."

The characteristics of an LCG are as follows:

1. The next random number depends on the previous random number.
2. The random numbers generated by an LCG have a modulo function that creates the appearance of random numbers.
3. The success of an LCG depends on the values of a, c, and m.
4. A lack of randomness may result if m is the power of 2.

Example 1 describes this method.

Example 1

Let us consider the following values for an LCG:

$$x_0(\text{seed}) = 25; \; a = 6; \; c = 37; \; m = 100$$

The sequence of random numbers is calculated as follows:

$$x_0 = 25$$

$$x_1 = (6 \times 25 + 37)\,\text{mod}\,100 = 87$$

$$R_1 = \frac{87}{100} = 0.87$$

$$x_2 = (6 \times 87 + 37)\,\text{mod}\,100 = 59$$

$$R_2 = \frac{59}{100} = 0.59$$

The sequence of random numbers is provided in Table 5.1.

From $i = 26$, the same sequence of random numbers starts again. Thus, the length of the cycle is 25. The target is to make the cycle a very long one. This depends on the values of x_0, a, c, and m. Based on the value of m, the cases considered are as follows [4]:

Case I. If $m = 2^b$ and $c \neq 0$, then the maximum length of the cycle (period) will be $P = m = 2^b$, provided c is prime to m and $a = 1 + 4k$, where k is an integer.

TABLE 5.1
Sequence of Random Numbers for Example 1

i	x_i	R_i
0	25	
1	87	0.87
2	59	0.59
3	91	0.91
4	83	0.83
5	35	0.35
6	47	0.47
7	19	0.19
8	51	0.51
9	43	0.43
10	95	0.95
11	7	0.07
12	79	0.79
13	11	0.11
14	3	0.03
15	55	0.55
16	67	0.67
17	39	0.39
18	71	0.71
19	63	0.63
20	15	0.15
21	27	0.27
22	99	0.99
23	31	0.31
24	23	0.23
25	75	0.75
26	87	0.87
27	59	0.59

Case II. If $m = 2^b$ and $c = 0$, then the maximum length of the cycle (period) will be $P = m/4 = 2^{b-2}$, provided $a = 3 + 8k$ or $a = 5 + 8k$, where k is an integer.
Case III. If m is a prime number and $c = 0$, then the maximum length of the cycle (period) will be $P = m - 1$, provided $a^k - 1$ is divisible by m, where $k = m - 1$.

Now, consider the following example:

Example 2

The values for the constants and seed are assumed as follows (Table 5.2):

$$x_0 = 25; \ m = 2^5; \ c = 13; \ a = 5 \ (k = 1)$$

TABLE 5.2

Sequence of Random Numbers for Example 2

i	x_i	R_i
0	25	
1	10	0.32
2	31	0.97
3	8	0.25
4	21	0.66
5	22	0.69
6	27	0.84
7	20	0.63
8	17	0.53
9	2	0.06
10	23	0.72
11	0	0

Example 2 clearly shows that if a zero appears somehow, then all the subsequent random numbers will be zero. This is one of the lacunas of LCGs that must be taken care of. However, a large value of m and a large value of a are required for an effective LCG. The seed value may be chosen based on a random event. Typically, it may be based on the date and time. The date and time may be entered in an expression that will return an integer, and after that, we need to be sure that it is an odd number.

5.3.2.2 Multiplicative Congruential Generator

A multiplicative congruential generator is just a modification of an LCG with the constant $c = 0$. Thus, the expression for the multiplicative congruential generator is

$$x_{n+1} = ax_n \bmod m \tag{5.3}$$

A numerical example based on the same values of the constants in Example 1 is given as follows:

Example 3

The values of the constants are same and the seed value is $x_0 = 23$. The cycle length, observed from Table 5.3, is 6.

5.3.2.3 Inversive Congruential Generator

The characteristics of an inversive congruential generator are as follows:

1. It is a non-LCG.
2. It uses a modular multiplicative inverse to generate random numbers.
3. The maximum period for this generator is q, where q is the modulo.

TABLE 5.3

Sequence of Random Numbers for Example 3

i	x_i	R_i
0	23	
1	38	0.38
2	28	0.28
3	68	0.68
4	8	0.08
5	48	0.48
6	88	0.88
7	28	0.28

The expression for this generator is as follows:

$$x_{n+1} = (ax_n^{-1} + c) \bmod q \text{ if } x_n \neq 0 \tag{5.4}$$

or

$$x_{n+1} = c \text{ if } x_n = 0 \tag{5.5}$$

5.3.2.4 Combined LCG

A combined LCG combines more than one LCG so as to get rid of the disadvantages of any single LCG. L'Ecuyer [5] described how such combinations can be done.

Let $Z_{i,1}, Z_{i,2}, \ldots, Z_{i,k}$ be some discrete random variables and one of them is a uniformly distributed random variable, distributed over $[0, m-2]$; then Z_i can be expressed as

$$Z_i = \left(\sum_{j=1}^{k} Z_{i,j} \right) \bmod m - 1 \tag{5.6}$$

where:

Z_i is uniformly distributed over $[0, m-2]$

This result can be used to form combined generators, such as

$$y_i = \left[\sum_{j=1}^{k} (-1)^{j-1} y_{i,j} \right] \bmod m - 1 \tag{5.7}$$

and

$$R_i = \begin{cases} \dfrac{y_i}{m}, & y_i > 0 \\ \dfrac{m-1}{m}, & y_i = 0 \end{cases} \tag{5.8}$$

5.3.2.5 Lagged Fibonacci Generator

A lagged Fibonacci generator has been developed based on the concept of Fibonacci series where the next number in the series is the summation of the previous two numbers. Thus, this method requires k past values. The expression for this generator is

$$x_n = (x_{n-j} \odot x_{n-k}) \bmod m, \ 0 < j < k \qquad (5.9)$$

where:

j and k are lags

$\odot$ is a binary operator such as $+, -, \times,$ or XOR

Example 4 can make this concept clearer.

Example 4

Assume the following values: $k = 2$ and $j = 1$, so that the expression for a lagged Fibonacci generator becomes $x_n = (x_{n-1} + x_{n-2}) \bmod m$, and thus the operator $\odot$ is the addition operator ($+$) and $m = 100$ is assumed. Also assume that $x_0 = 25$ and $x_1 = 31$.

Table 5.4 shows the random numbers generated by a lagged Fibonacci generator.

TABLE 5.4

Sequence of Random Numbers for Example 4

i	x_i	R_i
0	25	
1	31	
2	56	0.56
3	87	0.87
4	43	0.43
5	30	0.30
6	73	0.73
7	3	0.03
8	76	0.76
9	79	0.79
10	55	0.55
11	34	0.34
12	89	0.89
13	23	0.23
14	12	0.12
15	35	0.35
16	47	0.47
17	82	0.82
18	29	0.29

(Continued)

TABLE 5.4
(Continued) Sequence of Random Numbers
for Example 4

i	x_i	R_i
19	11	0.11
20	40	0.40
21	51	0.51
22	91	0.91
23	42	0.42
24	33	0.33
25	75	0.75
26	8	0.08
27	83	0.83
28	91	0.91
29	74	0.74
30	65	0.65
31	39	0.39
32	4	0.04
33	43	0.43
34	47	0.47
35	90	0.90
36	37	0.37
37	27	0.27
38	64	0.64
39	91	0.91
40	55	0.55
41	46	0.46

Thus, the lagged Fibonacci generator is an improvement over the LCG. This method is also computationally simple. The maximum period of this method may be $(2^j - 1) \times 2^{m-1}$.

5.3.2.6 Mid-Square Method

The mid-square method is also a very simple method. In this method, a seed value is taken first. Then, the square of this value is calculated. From the result obtained, the mid-value is again extracted and becomes the next random number. Example 5 illustrates this method.

Example 5

Suppose the seed value is $x_0 = 7612$.
    Square this value and thus we get $x_0^2 = 57\underline{9425}44$.
    The mid-value is 9425 and thus the next number is 9425.
    Thus, we have $R_1 = 0.9425$.

TABLE 5.5

Sequence of Random Numbers for Example 5

i	x_i	x_i^2	R_i
0	7612	57<u>9425</u>44	0.9425
1	9425	88<u>8306</u>25	0.8306
2	8306	68<u>9896</u>36	0.9896
3	9896	97<u>9308</u>16	0.9308
4	9308	86<u>6388</u>64	0.6388
5	6388	40<u>8065</u>44	0.8065
6	8065	65<u>0442</u>25	0.0442

Note: The underlined numbers indicate mid-values.

Table 5.5 shows some more values for this example.

The main disadvantage of this method is that if zeros appear as mid-digits, the subsequent numbers may become zero.

Pseudorandom number generators are mainly used in various computer software. Section 5.3.3 provides an overview of the software implementations of these methods.

5.3.3 SOFTWARE IMPLEMENTATION OF PSEUDORANDOM NUMBER GENERATORS

Every software package has a provision for generating random numbers. Generally, one of the pseudorandom number generators is used as the random number generator in these software.

The most common pseudorandom number generator in computer languages is the LCG. In addition, there are numerous other algorithms observed in various software packages. Some of these algorithms are as follows:

- Mersenne twister generator
- Marsaglia generator
- Linear feedback shift register
- Park–Miler random number generator
- Naor Reingold pseudorandom number generator
- Multiply-with-carry
- Complementary-multiply-with-carry

Among all these algorithms, the two very efficient algorithms, that is, the Mersenne twister generator and the Marsaglia generator, are discussed in Sections 5.3.3.1 and 5.3.3.2, respectively.

5.3.3.1 Mersenne Twister Generator

The Mersenne twister algorithm was originally developed by Makoto Matsumoto and Takuji Nishimura in 1997. The characteristics of this algorithm are as follows:

1. It is a pseudorandom number generator.
2. It is widely used in software such as MATLAB®, Python, C++, PHP, and Ruby.

3. It can generate random numbers with both 32-bit and 64-bit word lengths.
4. For a k-bit word length, it generates random numbers in the range $[0, 2^k - 1]$ over which the random numbers are distributed uniformly.
5. For a 32-bit word length, the length of one period is $2^{19937} - 1$.
6. The random numbers generated by this algorithm pass most of the statistical tests.
7. It can take a very long time to recover from poor initialization.
8. It uses a generalized feedback shift register.

This algorithm is based on a 32-bit word length and generates random numbers uniformly distributed over the range $[0, 2^{32} - 1]$. If the seed is given, the algorithm will be as provided in Figure 5.2.

For generating pseudorandom numbers based on an index value, the following algorithm is used (Figure 5.3). Let x denote the index number.

5.3.3.2 Marsaglia Generator

The Marsaglia generator was developed by George Marsaglia in 1988. The characteristics of this algorithm are as follows:

1. It is a combination of a lagged Fibonacci series and an arithmetic series.
2. It is a pseudorandom number generator.
3. The lagged Fibonacci series used for a Marsaglia generator has a period of length $(2^{24} - 1) \times 2^{94} \approx 2^{110}$.
4. It does not pass all the statistical tests.

The algorithm for a Marsaglia generator is given in Figure 5.4.

There are numerous other pseudorandom number generators used in various software packages. These generators are widely used in cryptography. There is a certain algorithm for every computer-generated random number; therefore, if the seed value and the algorithm are known, the random numbers can be susceptible to malicious attacks. Section 5.3.4 discusses the various attacks on pseudorandom number generators in computer networks.

1	Initialize MT[0] ← seed
2	Initialize i = 0
3	Calculate MT[i] ← last 32 bits of (1812433253 * (MT[i-1] XOR (right shift by 30 bits (MT[i-1])))+i)
4	Increment i
5	If i <623 Then
	Go to Step 3
	Else
	Go to Step 6
6	Stop

FIGURE 5.2 Mersenne twister algorithm when seed is given.

```
1    If x == 0 Then Go to Step 2
     Else Go to Step 8
     EndIf
2    Initialize i ← 0
3    Calculate y ← (MT[i] & 0x80000000) + MT[(i+1) mod 624] &
       0x7fffffff)
4    Calculate MT[i] ← MT[(i | 397) mod 624] XOR (right shift by 1 bit
       (y)]
5    If y mod 2 != 0 Then
         Calculate MT[i] ← MT[i] XOR (2567483615)
     EndIf
6    Increment i
7    If i < 624 Then
         Go to Step 3
     Else
         Go to Step 8
     EndIf
8    Assign y ← MT[x]
9    Calculate
         y ← y XOR (right shift by 11 bits (y))
         y ← y XOR (left shift by 7 bits (y) and (2636928640))
         y ← y XOR (left shift by 15 bits (y) and (4022730752))
         y ← y XOR (right shift by 18 bits (y))
10   Set x ← (x+1) mod 624
11   Stop
```

FIGURE 5.3 Mersenne twister algorithm when index is provided.

```
1  Take x_n from a Lagged Fibonacci Series
2  Take y_n from an Arithmetic Series
3  If x ≥ y Then
       The number is x – y
   Else
       The number is x   y | 1
   EndIf
```

FIGURE 5.4 Basic Marsaglia generator algorithm.

5.3.4 ATTACKS ON PSEUDORANDOM NUMBER GENERATORS

Pseudorandom number generators are used for various parameters in cryptography. Some of the parameters are parameters in digital signatures, session keys, hashing, and so on. If the pseudorandom number generators are intercepted, then the decoding of the messages becomes easy. These interceptions or attacks may come in various forms. The attacks may be classified as follows:

1. *Direct cryptanalytic attack.* This type of attack may take place when the attacker is unable to view the output of the generator. In this case, the attacker endeavors to analyze the encrypted message. This is the most common and frequent type of attack.
2. *Input-based attack.* In this type of attack, the attacker intends to control the input to analyze the output of the random number generator. Several possibilities for finding the input may arise. Examples are provided as follows:

a. There may be some inputs that apparently seem to be difficult to guess but turn out to be interpreted very easily. For example, the disk latency time, which seems to be difficult to get, may be tracked by an attacker.

b. The inputs for smartcards are fed to the pseudorandom number generators. These inputs may be intercepted by an attacker. In this case, the attacker may have to apply his/her own idea for getting the exact input.

c. Another type of input interception is possible wherein the attacker does not have enough of an idea for getting the exact input.

3. *State compromise extension attack.* This type of attack takes place when there is a weak hole in the security provision. An attacker finds out whether any previous attacks happened. If any did, the attacker finds that particular weak link and just extends the previous attack.

This type of attack is possible when the attacker knows the internal state of the system at any point in time. The attacker may try to find out the outputs from these known internal states. If there is success, it would indicate that the approach followed by the attacker is being fruitful. Such an attack is possible for a system that starts from a known state or a state that can be found out easily. These attacks can be further classified as follows:

a. Backtracking attacks are applicable when it is possible to find all the previous outputs from a single state.

b. Permanent compromise attacks are applicable when, from a particular state, all the previous output states as well as the future output states can be known.

c. Iterative guessing attacks are applicable when, from a particular state, all the future states can be known iteratively.

d. Meet-in-the-middle attacks are applicable when, from a particular state, all the previous outputs as well as the future outputs within a particular time frame can be found out. This is a mixture of an iterative guessing attack and a backtracking attack.

5.3.5 Quasi-Random Number Generators

The characteristics of quasi-random number generators are given as follows:

1. They generate highly uniform sampling.
2. Quasi-random numbers are actually nonrandom numbers.
3. The generated numbers have correlations among themselves.
4. These numbers may not pass many of the statistical tests.

Although the quasi-random numbers are not random numbers, they still have certain advantages over the pseudorandom number generators, which is why such numbers are in use and in some cases are even very successful. The main advantages of quasi-random number generators over pseudorandom number generators can be expressed through the following points:

1. For pseudorandom number generators, it has been observed that sometimes some portions of the entire domain are either oversampled or undersampled. But quasi-random number generators perform uniform sampling over the entire domain.
2. Pseudorandom numbers, in some cases, seem to be too random, which is never the case for quasi-random number generators.
3. Quasi-random number generators sometimes result in smaller errors than pseudorandom number generators.

There are some methods for generating quasi-random numbers. Some of the particularly significant methods are as follows:

1. Sobol's sequence
2. Gray code method

In a Sobol's sequence, the sequence uses some direction numbers. The jth number is obtained by performing XOR over all the direction numbers. In the Gray code method, the Gray code of a number is used instead of the actual number. The reason is that the successive Gray codes differ only by a one-bit position. Because of this fact, we can obtain the next quasi-random number just by an XOR operation.

5.4   TESTS FOR RANDOM NUMBER GENERATORS

There are several different types of tests for testing the various characteristics of generated random numbers by a random number generator. The different types of tests can be categorized into qualitative and quantitative tests. Since the quantitative tests are mostly used for testing the random number generators, the remaining part of this section discusses some of the frequently applied quantitative tests.

There are a significant number of quantitative tests for random numbers. A list of such tests is given as follows:

1. Frequency test
2. Runs test
3. Autocorrelation test
4. Poker test
5. Gap test
6. Lattice test
7. Spectral test

These are the most applied tests for random numbers. Some other tests include the following:

1. Equidistribution test
2. Coupon collectors test

   3. Permutation test
   4. Maximum of t test
   5. Birthday spacing test
   6. Collision test (hash test)
   7. Serial correlation test
   8. Serial test
   9. Blocking test
  10. Repeating time test
  11. Greatest common divisor (GCD) test
  12. Gorilla test
  13. Ising model test
  14. Random walk test
  15. N-block test
  16. Random walker on a line (S_n test)
  17. Two-dimensional intersection test
  18. Two-dimensional height correlation test
  19. Sum of independent distributions test
  20. Fourier transform test
  21. Universal statistical test
  22. Overlapping 5-permutation test
  23. Ranks of binary matrices test
  24. Bitstream test
  25. Parking lot test
  26. Count-the-1's test
  27. Overlapping sums test
  28. Squeeze test
  29. Minimum distance test
  30. Random sphere test
  31. Craps test

In addition, there are also some other tests such as visual tests and theoretical tests. But these tests are not described in this chapter because of the lack of automation for these tests. Some of these tests are described in Sections 5.4.1 through 5.4.9 in brief. First, we check for uniformity and independence of the generated random numbers. For testing the uniform distribution of the numbers, the following hypotheses are used:

   H_0: The numbers are distributed in the range [0,1] uniformly.
   H_1: The numbers are not distributed in the range [0,1] uniformly.

Here, H_0 and H_1 are the null and alternate hypothese, respectively.

   Similarly, for the independence property of the numbers generated, the null and alternate hypotheses are as follows:

   H_0: The generated numbers are independent.
   H_1: The generated numbers are not independent.

Suppose the level of significance is α and thus

$$\alpha = P(\text{reject } H_0 \mid H_0 \text{ is true})$$

Naturally, a type I error will occur if we reject a hypothesis on the basis of only one test, and a type II error will occur if we accept a hypothesis if only one test is successful. The tests are now discussed one after the other in Sections 5.4.1 through 5.4.9.

5.4.1 FREQUENCY TEST

The frequency test determines the frequency of the numbers over the range [0,1]. In other words, the frequency test determines whether the generated numbers follow a uniform distribution. This can be tested by either the chi-squared test or the Kolmogorov–Smirnov test. Both tests compare the continuous uniform distribution $F(x)$ with the observed distribution of the generated numbers.

The basic procedure for the Kolmogorov–Smirnov test is given through an example as follows: Let us take the following generated numbers:

$$0.43 \quad 0.64 \quad 0.31 \quad 0.04 \quad 0.81 \quad 0.25$$

Step 1: First, the generated numbers are arranged in ascending order. Thus, the ordered numbers are as follows:

$$0.04 \quad 0.25 \quad 0.31 \quad 0.43 \quad 0.64 \quad 0.81$$

Step 2: Calculate the following:

$$D^+ = \max\left(\frac{i}{N} - R_i\right)$$

and

$$D^- = \max\left(R_i - \frac{i-1}{N}\right)$$

where:
   N is the number of generated numbers
   R_i is the ith generated number

Thus, the values of D^+ and D^- are calculated as follows:

$$D^+ = \max\left[\left(\frac{1}{6} - 0.04\right), \left(\frac{2}{6} - 0.25\right), \left(\frac{3}{6} - 0.31\right), \left(\frac{4}{6} - 0.43\right), \left(\frac{5}{6} - 0.64\right), \left(\frac{6}{6} - 0.81\right)\right]$$

$$= \max(0.13, 0.08, 0.17, 0.24, 0.19, 0.19)$$

$$= 0.24$$

$$D^- = \max\left[\left(0.04 - \frac{0}{6}\right), \left(0.25 - \frac{1}{6}\right), \left(0.31 - \frac{2}{6}\right), \left(0.43 - \frac{3}{6}\right), \left(0.64 - \frac{4}{6}\right), \left(0.81 - \frac{5}{6}\right)\right]$$

$$= \max(0.04, 0.08, -0.02, -0.07, -0.03, -0.02)$$

$$= 0.08$$

Step 3: Find out $D = \max(D^+, D^-)$
Thus, we have

$$D = \max(0.24, 0.08) = 0.24$$

Thus, the observed value is 0.24.

Step 4: Now, for given values of N and α, Kolmogorov–Smirnov critical values can be obtained. For this example, we have $N = 6$, and the critical values are given in the following table for different values of α:

α	Critical Value
0.01	0.618
0.05	0.521
0.10	0.470

Step 5: If the observed value (D) is less than or equal to D_α, we conclude that there is no difference between the true distribution and the uniform distribution. Otherwise, the null hypothesis is rejected.

For the current example, $D = 0.24$, which is less than the critical values for all the α's as shown in the table above. Thus, we can conclude that for any value of α, there is no difference between the observed distribution and the uniform distribution.

The chi-squared test can also be applied for the same purpose. We know that the chi-squared test uses the following statistic:

$$\chi^2 = \sum_{i=1}^{n} \frac{(\text{Observed}_i - \text{Expected}_i)^2}{\text{Expected}_i} \tag{5.10}$$

where:
  Observed$_i$ and Expected$_i$ are the observed and expected values, respectively, of the ith class

Expected$_i$ is given by

$$\text{Expected}_i = \frac{N}{n}$$

where:
  N is the total number of observations made
  n is the number of classes

5.4.2 RUNS TEST

The runs test tests the arrangement of the generated numbers to check whether the numbers are independent. Here, we count the number of runs up and runs down. If the number is increased by a value, we call it "runs up"; otherwise we call it "runs down."

For example, consider the following numbers:

$$0.62 \quad 0.92 \quad 0.28 \quad 0.63 \quad 0.47 \quad 0.10 \quad 0.43 \quad 0.58 \quad 0.24 \quad 0.15$$

$$0.56 \quad 0.65 \quad 0.21 \quad 0.38 \quad 0.32 \quad 0.29 \quad 0.93 \quad 0.64 \quad 0.89 \quad 0.74$$

$$0.51 \quad 0.96 \quad 0.52 \quad 0.94 \quad 0.14 \quad 0.49 \quad 0.26 \quad 0.18 \quad 0.34 \quad 0.13$$

$$0.82 \quad 0.33 \quad 0.68 \quad 0.17 \quad 0.57 \quad 0.73 \quad 0.69 \quad 0.88 \quad 0.48 \quad 0.79$$

We have to find out whether the hypothesis of independence can be rejected for $\alpha = 0.05$.

For this example, we will have to find out the runs up and the runs down first. The sequence of runs up and runs down are given below.

$$+ \; - \; + \; - \; - \; + \; + \; - \; - \; + \; + \; - \; + \; - \; - \; + \; - \; + \; -$$

$$- \; + \; - \; + \; - \; + \; - \; - \; + \; - \; + \; - \; + \; - \; + \; + \; - \; + \; - \; +$$

There are 31 runs observed in the above sequence, and thus we represent this number by a so that $a = 31$ (a: total number of runs as observed). The total number of observations is 40 and thus $N = 40$.

The mean and variance of a are given as follows:

$$\mu_a = \frac{2N-1}{3}$$

$$\sigma_a = \frac{16N-29}{90}$$

Thus, for this example we have

$$\mu_a = \frac{2 \times 40 - 1}{3} = 26.33$$

$$\sigma_a = \frac{16 \times 40 - 29}{90} = 6.79$$

Thus, by the normal distribution formula, we now find

$$Z_0 = \frac{31 - 26.33}{\sqrt{6.79}} = 1.79$$

The hypothesis of independence fails if the following condition fails:

$$-Z_{\alpha/2} \leq Z_0 \leq Z_{\alpha/2}$$

If the level of significance $\alpha = 0.05$, then $\alpha/2 = 0.025$.

Now from the "area under standard normal distribution curve table," we get $Z_{0.025} = 1.96$. Therefore, the calculated value $Z_0 = 1.79$ is less than $Z_{0.025} = 1.96$; thus, the hypothesis of independence is not rejected.

5.4.3 Autocorrelation Test

The autocorrelation test tests the correlation among the generated numbers and compares the calculated correlation with the expected correlation of zero. This test is also being explained through an example.

Consider the following set of generated numbers:

$$0.34 \quad 0.27 \quad 0.58 \quad 0.66 \quad 0.14 \quad 0.71 \quad 0.49 \quad 0.31 \quad 0.16 \quad 0.86$$

$$0.56 \quad 0.37 \quad 0.48 \quad 0.27 \quad 0.82 \quad 0.07 \quad 0.43 \quad 0.39 \quad 0.15 \quad 0.23$$

$$0.93 \quad 0.62 \quad 0.54 \quad 0.24 \quad 0.38 \quad 0.40 \quad 0.79 \quad 0.57 \quad 0.29 \quad 0.36$$

To find the autocorrelation, we find the following statistic:

$$Z_0 = \frac{\rho_{im}}{\sigma_{\rho im}}$$

where

$$\rho_{im} = \frac{1}{M+1}\left[\sum_{k=0}^{M} R_{i+km}R_{i+(k+1)m}\right] - 0.25$$

where:

i is the first number to test in the sequence

m is the lag

M is an integer such that it satisfies the expression $i + (M+1)m \leq N$

$$\sigma_{\rho im} = \frac{\sqrt{13M+7}}{12(M+1)}$$

The hypotheses are given by

H_0: ρ_{im} is zero.

H_1: ρ_{im} is not zero.

If the condition $-Z_{\alpha/2} \leq Z_0 \leq Z_{\alpha/2}$ is satisfied, accept the null hypothesis.

Suppose now that we want to find the correlation among the second, sixth, and eighth numbers.

Since we will have to start with the second number, $i = 2$. The value of m is $m = 4$ since we want to find the correlation among every fourth number. In the sample above, there are 30 numbers and thus $N = 30$. Thus, the value of M is $M = 6$.

Hence, the value of ρ_{im} is calculated as

$$\rho_{im} = \frac{1}{6+1}\left[R_2R_6 + R_6R_{10} + R_{10}R_{14} + R_{14}R_{18} + R_{18}R_{22} + R_{22}R_{26} + R_{26}R_{30}\right] - 0.25$$

$$= \frac{1}{7}\left[\begin{array}{c} 0.27 \times 0.71 + 0.71 \times 0.86 + 0.86 \times 0.27 + 0.27 \times 0.39 \\ + 0.39 \times 0.62 + 0.62 \times 0.40 + 0.40 \times 0.36 \end{array}\right] - 0.25$$

$$\approx 0.0034$$

The value of $\sigma_{\rho im}$ is calculated as

$$\sigma_{\rho im} = \frac{\sqrt{13 \times 6 + 7}}{12(6+1)} = \frac{9.2195}{84} = 0.1098$$

Thus, Z_0 can be calculated as

$$Z_0 = \frac{0.0034}{0.1098} = 0.031$$

Now if the level of significance is $\alpha = 0.05$, then $Z_{\alpha/2} = Z_{0.025} = 1.96$.

Thus, the value of $Z_0 = 0.031$ satisfies the condition $-Z_{\alpha/2} \leq Z_0 \leq Z_{\alpha/2}$, and therefore the null hypothesis is accepted on the basis of this particular test.

5.4.4 POKER TEST

The poker test checks whether the same digit appears more than once in the generated random numbers, and if the digits are repeated in the numbers, it checks whether the respective numbers are independent. Consider the following series of numbers. In the example of 10 numbers shown below, a certain digit is repeated in each number. For example, digit 5 is repeated twice; in the third number, digit 3 is repeated thrice:

0.255 0.772 0.333 0.656 0.331 0.454 0.818 0.344 0.292 0.121

However, in a three-digit number, there are three possibilities of repetition: (1) the probability of having three different digits, (2) the probability of having two of the same digits, and (3) the probability of having three of the same digits. These probabilities can be calculated as follows:

1. Probability of having three different digits

$$P\,(\text{three different digits}) = (10 \times 9 \times 8)/(10 \times 10 \times 10)$$

$$= (9 \times 8)/(10 \times 10) = 0.9 \times 0.8 = 0.72$$

2. Probability of having two of the same digits

$$P(\text{two same digits}) = (0.1) \times (0.1) = 0.01$$

TABLE 5.6
Calculations for Chi-Squared Test

Possibilities	Observed Frequency	Expected Frequency	$\dfrac{(\text{Observed}_i - \text{Expected}_i)^2}{\text{Expected}_i}$
Three different digits	620	$0.72 \times 1000 = 720$	$\dfrac{(620-720)^2}{720} = 13.89$
Three same digits	40	$0.01 \times 1000 = 10$	$\dfrac{(40-10)^2}{10} = 90$
Two same digits	340	$0.27 \times 1000 = 270$	$\dfrac{(340-270)^2}{270} = 18.15$
Total	1000	1000	122.04

Suppose a total of 1000 three-digit numbers are generated, and among them 620 are numbers with different digits, 40 are numbers with the same digits, and 340 are numbers with two of the same digits. The calculations for the basic chi-squared test are given in Table 5.6.

If the confidence level is $\alpha = 0.05$, the value of the respective χ^2 with degree of freedom 2 is given by $\chi^2 = 5.99$ (found from Table 5.6). Since the value $\chi^2 = 5.99 < 122.04$, the numbers are not at all independent. Thus, the hypothesis that the "numbers are independent" is strongly rejected.

5.4.5 GAP TEST

The gap test measures the "gap" between the occurrences of the same number in a series of generated random numbers. Consider the following series of numbers:

$$4, 1, 3, 5, 1, 7, 2, 8, 2, 0, 7, 9, 1, 3, 5, 2, 7, 9, 4, 1, 6, 3$$

$$3, 9, 6, 3, 4, 8, 2, 3, 1, 9, 4, 4, 6, 8, 4, 1, 3, 8, 9, 5, 5, 7$$

$$3, 9, 5, 9, 8, 5, 3, 2, 2, 3, 7, 4, 7, 0, 3, 6, 3, 5, 9, 9, 5, 5$$

$$5, 0, 4, 6, 8, 0, 4, 7, 0, 3, 3, 0, 9, 5, 7, 9, 5, 1, 6, 6, 3, 8$$

$$8, 8, 9, 2, 9, 1, 8, 5, 4, 4, 5, 0, 2, 3, 9, 7, 1, 2, 0, 3, 6, 3$$

Suppose we want to measure the gap between the successive occurrences of the digit 3. The total number of 3s in the above sequence is 18. The gap between the successive occurrences of the digit 3 is given in Table 5.7.

The purpose is to find the frequency of the gaps. If all the digits have an equal chance of occurring, the probability of getting a 3 is $1/10 = 0.1$. Thus, the probability

TABLE 5.7

Gap between the Successive Occurrences of the Digit 3

	Gap Length
Gap between the 1st and 2nd occurrences	10
Gap between the 2nd and 3rd occurrences	7
Gap between the 3rd and 4th occurrences	0
Gap between the 4th and 5th occurrences	2
Gap between the 5th and 6th occurrences	3
Gap between the 6th and 7th occurrences	8
Gap between the 7th and 8th occurrences	5
Gap between the 8th and 9th occurrences	5
Gap between the 9th and 10th occurrences	2
Gap between the 10th and 11th occurrences	4
Gap between the 11th and 12th occurrences	1
Gap between the 12th and 13th occurrences	14
Gap between the 13th and 14th occurrences	0
Gap between the 14th and 15th occurrences	9
Gap between the 15th and 16th occurrences	14
Gap between the 16th and 17th occurrences	5
Gap between the 17th and 18th occurrences	1

of not getting 3 is $1 - 0.1 = 0.9$. Thus, the probability of a gap of 10 is $(0.9)^{10}(0.1)$. Thus, the frequency distribution of all the digits will be given by

$$\sum_{i=0}^{x} (0.9)^i (0.1)$$

Now for all the digits, the number of gaps is given in Table 5.8.

On the above data, either the chi-squared test or the Kolmogorov–Smirnov test can be applied in the same way as shown earlier.

TABLE 5.8

Number of Gaps for All Digits

Digit	Number of Gaps
0	7
1	8
2	8
3	17
4	10
5	13
6	7
7	8
8	9
9	13

5.4.6 Equidistribution Test

The equidistribution test determines whether the generated random numbers are equally distributed. The steps for this test are as follows:

1. The total N random numbers in the interval (a,b) are divided into k classes $n_1, n_2,..., n_k$ with sizes $N_1, N_2,..., N_k$, where $N_1 + N_2 + \cdots + N_k = N$.
2. For each class n_i, the expected number is calculated. Each of these numbers is assumed to have the same probability $p = (b - a)/k \times N$.
3. To check the probability for the classes, apply the chi-squared test and apply the Kolmogorov–Smirnov test to test the entire data set.

5.4.7 Coupon Collectors Test

In the coupon collectors test, the sequence of generated random numbers is scanned and the lengths of successive "complete sets" of integers between 0 and $d - 1$ are recorded. Suppose that the sequence of random numbers is represented by $X_1, X_2,...$ and the subsequence is represented by $X_{i+1}, X_{i+2},..., X_{i+r}$. The length of r of the subsequence is first counted to obtain the complete sets of integers from 0 to $d - 1$. The minimum length of r is d and the maximum length is not fixed or bounded. Thus, a value t is defined as the upper bound and the range of r is $d \leq r < t$.

The test is run until n complete sets of integers between 0 and $d - 1$ are obtained. In each case, the length r is recorded where $d \leq r < t$. All the other sequences that are longer than t are also gathered separately. Then, the chi-squared test can be performed with $t - d + 1$ degrees of freedom. The expected probabilities for each type of length are given by

$$P_r = \frac{d!}{d^r}(^{r-1}C_{d-1}) \text{ for } d \leq r < t \tag{5.11}$$

and

$$P_t = 1 - \frac{d!}{d^{t-1}}(^{t-1}C_d) \tag{5.12}$$

The term $^{r}C_d$ is known as Stirling's number of the second kind.

5.4.8 Permutation Test

The permutation test is described as follows:

- The generated n random numbers are divided into g groups. Each group has p random numbers.
- The numbers in each group g can have $p!$ permutations or ordering. Count the occurrences of each possible ordering in the group.
- Perform a chi-squared test with $k = t!$ degrees of freedom and with probability $1/t!$ on each of the permutations.

5.4.9 MAXIMUM OF t TEST

The procedure for the maximum of t test is as follows:

- The generated n random numbers are divided into g groups. Each group has p random numbers.
- Find the maximum random number in each group. Thus, we get a sequence u of the maximum numbers of the groups.
- The generated sequence of maximum numbers of g groups is subjected to the Kolmogorov–Smirnov test. The sequence of maximum numbers is assumed to have a distribution $F(x) = x^t$ for $0 \leq x \leq 1$.
- Next, k equidistant bins between [0,1] can be made. The probability of the lower bin is subtracted from the probability of the actual bin to obtain the expected number of values in each bin.
- The percentage in bin $i = 1,2,3,\ldots,k$ is given by $(i/k)^p - (i-1/k)^p$.
- The expected value can be obtained by multiplying the value per bin with the number of groups g.

5.5 CONCLUSION

In this chapter, the concept of random number generators is introduced. Various pure and pseudorandom number generators that are commonly used are described here. Some famous algorithms for generating random numbers are described. The concept of quasi-random numbers and various kinds of tests performed to test the various aspects of random number are introduced. The user is expected to get a good grasp of the concept of random number generators after going through this chapter.

REFERENCE

1. Learmonth, G. P. (1976). *Theory and Testing of Uniform Random Number Generators.* Monterey, CA: Naval Postgraduate School.

6 Random Variate Generation

6.1 INTRODUCTION

In Chapter 5, we described various probability distributions. This chapter assumes that the reader has a good grasp of all the basic and commonly used probability distributions discussed in Chapter 5. This chapter provides methods for sampling various discrete and continuous probability distributions. The method of generating sampling values from various probability distributions is known as the generation of random variates. This basically generates the random numbers that follow a particular probability distribution. These random numbers are generated using the cumulative distribution function or the density function. Therefore, for each probability distribution, the probability density function along with the cumulative distribution function must be known in advance.

To generate the random number for all these probability distributions, a steady source of random numbers is required. These random numbers can be taken from a source that generates uniformly distributed random numbers. There are several good existing algorithms that can be used for this purpose. For example, the Mersenne twister algorithm is used in several software packages, and this random number generation technique produces a significantly almost unbiased flow of random numbers. The generated random numbers can be used as the input for random variate generation.

The methods discussed in this chapter are the inverse transform technique, the convolution method, and the acceptance–rejection method. Although all software packages nowadays have built-in routines to generate random numbers, every software uses one or more particular type of probability distribution. However, a particular software cannot consider all the common types of probability distributions, and therefore, the method of generating random variates is required for practical use of these random number generators.

This chapter uses a steady source of random numbers for showing examples for every method. The random numbers have been generated by using the Mersenne twister algorithm in MATLAB® software since this algorithm seems to be a good source of pseudorandom numbers. We assume that the random numbers generated are uniformly distributed.

The remaining sections of this chapter are arranged as follows: Section 6.2 describes the various methods of random variate generation. Section 6.2.11 especially describes the inverse transform technique. Section 6.2.12 gives an overview of the convolution method. Section 6.2.13 depicts the acceptance–rejection method. Section 6.3 provides the conclusion for this chapter.

6.2 VARIOUS METHODS OF RANDOM VARIATE GENERATION

The distinction between random number generation and random variate generation is very subtle. However, there are various random variate generation methods depending on whether it is a discrete or continuous probability distribution. For a discrete probability distribution, the interval [0,1) is divided into n divisions, a pseudorandom number is generated, and a search is performed for the corresponding division or subinterval. This search can be performed in any of the following methods:

- Linear search
- Binary search
- Indexed search

Although there are various other search techniques, these three are the very commonly used basic techniques.

Each of the above-mentioned algorithms is described in Sections 6.2.1 through 6.2.3.

6.2.1 LINEAR SEARCH

The basic algorithm for this method is shown in Figure 6.1. Let the generated random number from the source generating uniformly distributed random numbers be X.

From the figure, it is clear that the method first generates a random number X from a random number generator that generates uniformly distributed random numbers. Then, a function is decided which scales the generated random number X to the distribution defined by the function f. After this, the series of interval subdivisions for the interval [0,1) are scanned. If the random number X falls within a certain interval, then the interval I is marked, and from this number, the function $f(I)$ generates the required random number.

6.2.2 BINARY SEARCH

As the name suggests, the binary search algorithm basically continues dividing the available interval or subinterval into two equal divisions, and the range of values for each division is checked to see whether the generated random number falls within the range. Naturally, the series to search for should be sorted in ascending

```
Scan the number of intervals I from either side of the overall interval [0, 1)
If X falls within interval I Then
        Mark the interval I
        Calculate f(I) where f is a previously defined function
Else
        Increment I = I + 1
EndIf
```

FIGURE 6.1 Algorithm for linear search.

> **Set** min $= 0$ **and** max $= size(I)$
>
> **While** max $\geq$ min **Do**
>
> **Calculate** mid $= midpo\mathrm{int}(\mathrm{min},\mathrm{max})$
>
> **If** $I(mid) < X$ **Then**
>
>   **Increase** min $= mid + 1$
>
>   **Else If** $I(mid) > X$ **Then**
>
>   **Decrease** max $= mid - 1$
>
>   **Else**
>
>   **Return** *mid*
>
> **End While**

FIGURE 6.2 Algorithm for binary search.

or descending order of values. Since this algorithm is used here for scanning the interval divisions in the range [0,1), the intervals are already ordered in ascending order of values by default. After identifying the particular subinterval where the generated random number can lie, the interval number I is marked as usual and the function $f(I)$ generates the required random variates following the distribution of $f()$.

The binary search algorithm is provided in Figure 6.2. The inputs to the algorithm are the generated random number X generated from a random number generator that generates uniformly distributed random numbers and the ordered list of divided intervals I—a minimum value, *min* (the value of *min* is usually 0), and a maximum value, *max* (the value of *max* is usually the total number of subintervals).

The binary search algorithm is a very efficient algorithm that searches for a particular item in $O(\log n)$ time. After searching for the particular interval i, the function $f(i)$ is applied over the selected interval number to generate the random number following the distribution of $f(i)$.

6.2.3 INDEXED SEARCH

An indexed search is similar to a linear search, although it is much more efficient than the linear search algorithm. An index for the intervals is maintained here, and the indexes are searched instead of the actual intervals. If, for some reason, the intervals are not ordered, then this search algorithm is very effective for faster search. After identifying a particular index, the respective interval is accessed directly to find the interval number. Then, the function $f(i)$ is applied over the selected interval i to generate a random number following the distribution of $f(i)$.

There are various other methods in the existing literature for random variate generation. The prime methods among them are as follows:

- Inverse transform technique
- Convolution method
- Acceptance–rejection method

These methods will be discussed in detail in Sections 6.2.11 through 6.2.13. In addition, there are also other methods for random variate generation, some of which include the following:

- Slice sampling
- Ziggurat algorithm
- Markov chain Monte Carlo (MCMC) method
- Metropolis–Hastings algorithm
- Gibbs sampling
- Box–Muller transform method
- Marsaglia polar method

These algorithms are briefly described in Sections 6.2.4 through 6.2.10.

6.2.4 SLICE SAMPLING

Slice sampling is a pseudorandom number sampling method. This is a type of MCMC algorithm. The method is depicted in Figure 6.3.

In this algorithm, first we choose a random curve and then slice the curve horizontally. A slice is then selected randomly from among them. We can sample a point within the curve by randomly selecting a slice that falls at or below the curve at the x-position from the previous iteration and then randomly pick an x-position somewhere along the slice. By using the x-position from the previous iteration of the algorithm, in the long run we select slices with probabilities proportional to the lengths of their segments within the curve.

Slice sampling can be applied to any randomly chosen curve. Slice sampling is the same as Gibbs sampling or the Metropolis method in terms of its purpose.

1	Generate a uniformly distributed random number x
2	Choose a random curve represented by $f(x)$
3	Choose an initial value x_0 for which $f(x_0) > 0$
4	Sample a value y between 0 and $f(x_0)$
5	Draw a horizontal line across the curve at y
6	Sample a point (x, y) from the line segment within the curve
7	Repeat from step 2 with a new x value

FIGURE 6.3   Method of slice sampling.

6.2.5 ZIGGURAT ALGORITHM

The Ziggurat algorithm is a type of pseudorandom number sampling method. It is also a type of rejection sampling method, which will be discussed in Section 6.2.13 in this chapter. The algorithm depends on the random number generated from a uniformly distributed random number generator source and precomputed tables. This algorithm is particularly applicable to monotonically decreasing probability distributions or symmetric unimodal distributions. It is conceptually based on covering the probability distribution with rectangular segments stacked in decreasing order of size, resulting in a figure that resembles a Ziggurat. The Ziggurat algorithm is faster than the Marsaglia polar algorithm and the Box–Muller transform method.

The Ziggurat algorithm generates a random point in a distribution, which is a little greater than the desired distribution. After this, the algorithm checks whether the generated random point is within the desired distribution. If it is not within the distribution, the method repeats itself. If the random point is within the probability distribution curve, then its x-coordinate is a random number with the desired distribution. This is the overall procedure of the Ziggurat algorithm.

The input to the Ziggurat algorithm is a monotonically decreasing probability distribution function $f(x)$, $\forall x \geq 0$. The base of the Ziggurat is defined as all points inside the distribution and below $y_1 = f(x_1)$. This consists of a rectangular region from $(0,0)$ to (x_1, y_1) and the (typically infinite) tail of the distribution, where $x > x_1$ (and $y < y_1$). This is called a layer, and this layer (layer 0) has area A. On top of this, add a rectangular layer of width x_1 and height A/x_1, so it also has the area A. The top of this layer is at height $y_2 = y_1 + A/x_1$ and intersects the distribution function at a point (x_2, y_2), where $y_2 = f(x_2)$. This layer includes every point in the distribution function between y_1 and y_2, but (unlike the base layer) also includes points such as (x_1, y_2) which are not in the desired distribution. The Ziggurat algorithm is described in Figure 6.4.

1	**Generate 2 random numbers** $U_0 \in [0,1)$ **and** $U_1 \in [0,1)$
2	**Choose a layer** i **randomly such that** $0 \leq i < n$
3	**Assign** $x = U_0 x_i$
4	**If** $x < x_{i+1}$ **Then Return** x
5	**Calculate** $y = y_i + U_1(y_{i+1} - y_i)$
6	**Compute** $f(x)$
7	**If** $y < f(x)$ **Then**
8	**Return** x
9	**Else**
10	**Choose new** $U_0 \in [0,1)$ **and** $U_1 \in [0,1)$
11	**Go to step 2**

FIGURE 6.4 Ziggurat algorithm.

6.2.6 MCMC Method

The Markov chain will be described in brief in Chapter 7. The MCMC method actually constructs a Markov chain based on a particular probability distribution. After a large number of steps, the equilibrium state is used as a sample from the desired probability distribution. The most difficult factor for this method is to determine how many steps are required to reach equilibrium within an acceptable error. The typical application for this method is to calculate multidimensional integrals. This method is widely used in computational physics, computational biology, Bayesian analysis, and computational linguistics.

The MCMC method also includes the random walk Monte Carlo method. In this method, it is assumed that a walker walks randomly. At each step of the walker, the integrand is calculated with the value represented by the step. Then, the walker makes a number of tentative steps to approximate the integrand toward accuracy and accordingly determines the accurate spot or value. A Markov chain is constructed to have the integrand as its equilibrium distribution. Some of the most significant random walk algorithms are slice sampling, Gibbs sampling, the Metropolis–Hastings algorithm, and a multiple-try Metropolis.

6.2.7 Metropolis–Hastings Algorithm

The Metropolis–Hastings algorithm is named after Nicholas Metropolis and W. K. Hastings. Metropolis proposed this algorithm in 1953, and the algorithm was later modified by Hastings in 1970 by generalizing the algorithm so as to make it applicable to various general cases. This sampling method is applicable to those probability distributions for which no direct sampling method can be applied. This is a type of MCMC method that is used to generate a sequence of random samples from a given probability distribution. This algorithm is generally applied to multidimensional distributions.

The Metropolis–Hastings algorithm is said to be applicable to any probability distribution $P(x)$ if $f(x)$ is proportional to the density of P. The algorithm is shown in Figure 6.5.

6.2.8 Gibbs Sampling

Gibbs sampling was proposed by Josiah Willard Gibbs for the theories of statistical physics. It is also a type of MCMC algorithm. Gibbs sampling is applicable to multivariate probability distributions. This sampling procedure is used when no direct method can be used for multivariate probability distributions. The Gibbs algorithm uses Bayesian inference. This sampling method can also be thought of as a special edition of the Metropolis–Hastings algorithm. The basic method for this algorithm is very simple. The method uses the concept of integration of the joint distribution of the variables. The detailed method of Gibbs sampling is not shown because of its hardcore mathematical background, which may not be easy to understand.

---

1 Choose a function $f(x)$ which is proportional to the probability distribution $P(x)$

2 Choose the first sample x_0 randomly

3 Choose a symmetric probability density function $Q(x|y)$ (also called proposal density or jumping distribution) in order to generate x based on the previous sample value y. For the function $Q(x|y)$ to be symmetric, $Q(x|y) = Q(y|x)$

4 Generate x from the proposal density $Q(x^*|x_t)$ where t is the iteration number

5 Calculate acceptance ratio $\alpha = \dfrac{f(x^*)}{f(x_t)}$

6 If $\alpha = \dfrac{f(x^*)}{f(x_t)} = \dfrac{P(x^*)}{P(x_t)}$ Then

7   f is proportional to P

8 Else

9   f is not proportional to P

10 If $\alpha \geq 1$ Then

11   Accept the candidate by setting $x_{t+1} = x^*$ or accept the candidate with probability α

12 Else

13   Reject the candidate

14   Set $x_{t+1} = x_t$

FIGURE 6.5 Metropolis–Hastings algorithm.

6.2.9 BOX–MULLER TRANSFORM METHOD

The Box–Muller transform method was proposed by George Edward Pelham Box and Mervin Edgar Muller in 1958. This method was proposed as an alternative method for the inverse transform technique. It can generate a pair of independent, standard normally distributed random numbers from a source of uniformly distributed random numbers. There are two forms of this method— the basic form and the polar form, which are described in Sections 6.2.9.1 and 6.2.9.2, respectively.

6.2.9.1 Basic Form

Suppose that $U_0 \in [0,1)$ and $U_1 \in [0,1)$ are the two random numbers generated from a source of uniformly distributed random numbers. Then, the random numbers Z_0 and Z_1 following a standard normal distribution are given by the following:

$$Z_0 = R\cos\theta = \sqrt{-2\ln U_1}\cos(2\pi U_2) \tag{6.1}$$

and

$$Z_1 = R\sin\theta = \sqrt{-2\ln U_1}\sin(2\pi U_2) \tag{6.2}$$

where

$$R^2 = -2\ln U_1 \text{ and } \theta = 2\pi U_2 \tag{6.3}$$

Equations 6.1 and 6.2 are based on the assumption that if $U_0 \in [0,1)$ and $U_1 \in [0,1)$ are two standard normally distributed numbers, then R^2 and θ will also follow standard normal distribution.

6.2.9.2   Polar Form

The polar form is based on the polar coordinates. If (X,Y) are Cartesian coordinates, then (R,θ) can be their polar coordinates, where R is the radius of the circle in polar form and θ is the angle of this radius with the center. This concept is used in finding the polar form of Equations 6.1 and 6.2. From Figure 6.6, it can easily be shown that

$$s = R^2 = u^2 + v^2 \tag{6.4}$$

Thus, we have

$$R = \sqrt{s} \tag{6.5}$$

The two values of Z_0 and Z_1, in polar form, can be given by

$$Z_0 = \sqrt{-2\ln U_1}\cos(2\pi U_2) = \sqrt{-2\ln s}\,\frac{u}{\sqrt{s}} = u\sqrt{\frac{-2\ln s}{s}} \tag{6.6}$$

and

$$Z_1 = \sqrt{-2\ln U_1}\sin(2\pi U_2) = \sqrt{-2\ln s}\,\frac{v}{\sqrt{s}} = v\sqrt{\frac{-2\ln s}{s}} \tag{6.7}$$

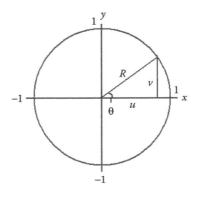

FIGURE 6.6   Polar coordinates.

6.2.10 Marsaglia Polar Method

The Marsaglia polar method is similar to the Box–Muller method but this algorithm is superior. This method also uses polar coordinates. Let (x,y) be the random points for $-1 < x < 1$ and $-1 < y < 1$, and $s = x^2 + y^2 < 1$. From these, the two random numbers generated will be given by the following equations:

$$Z_0 = x\sqrt{\frac{-2\ln s}{s}} \tag{6.8}$$

and

$$Z_1 = y\sqrt{\frac{-2\ln s}{s}} \tag{6.9}$$

In addition, some other methods are also available. The most commonly used methods for generating random variates are as follows: the inverse transform technique, the convolution method, and the acceptance–rejection method. Thus, Sections 6.3 through 6.5 are dedicated to each of these methods.

6.2.11 Inverse Transform Technique

The inverse transform technique makes the simple inverse of a function [1]. Naturally, this technique is useful for those distributions that are not difficult to create inversely. This technique can be used to get an inverse form from an exponential distribution, a Weibull distribution, a triangular distribution, and an empirical distribution. The basic principle of this technique is as follows:

Given the expression $F(X) = R$, we will have to find $X = F^{-1}(R)$. This means that if R is expressed in terms of X, then we will have to find X in terms of R. That is why it is called the inverse transform technique. The prerequisite for such technique is that the user must have knowledge of the expression for the underlying probability density function or the probability mass function.

The steps of the inverse transform technique are as follows:

1. Get the probability density function.
2. Compute the cumulative distribution function for the random variable X.
3. Set the relation $F(X) = R$.
4. Find the relation $X = F^{-1}(R)$ from $F(X) = R$, that is, find the expression for X in terms of R.
5. Generate the random numbers $R_1, R_2, \ldots$ from the source of uniformly generating random numbers.
6. Find the values of the random numbers $X_1, X_2, \ldots$ using the expression $X = F^{-1}(R)$ from the generated random numbers $R_1, R_2, \ldots$.

Let us now apply this method to some probability distributions as shown in Sections 6.2.11.1 through 6.2.11.4.

6.2.11.1 Exponential Distribution

The probability density function for the exponential distribution is given by

$$f(x) = \begin{cases} \lambda e^{-\lambda x}, & x \geq 0 \\ 0, & x < 0 \end{cases} \tag{6.10}$$

where:

λ is the average number of occurrences

It may be thought of as the arrival rate for the event of part arrivals, for example. Thus, $1/\lambda$ may be thought of as the average interarrival time. The cumulative distribution of the above probability density function can be calculated as follows:

$$F(x) = \int_{-\infty}^{x} f(t)dt$$

$$= \int_{-\infty}^{0} f(t)dt + \int_{0}^{x} f(t)dt$$

$$= 0 + \int_{0}^{x} f(t)dt \tag{6.11}$$

$$= \int_{0}^{x} \lambda e^{-\lambda t}dt$$

$$= \lambda \frac{e^{-\lambda t}}{-\lambda} |_{0}^{x}$$

$$= 1 - e^{-\lambda x}$$

Next, we set $F(X) = 1 - e^{-\lambda x} = R$ and find the expression for x in terms of R. Thus, we have the following:

$$R = 1 - e^{-\lambda x}$$

$$\Rightarrow e^{-\lambda x} = 1 - R$$

$$\Rightarrow \ln(e^{-\lambda x}) = \ln(1 - R) \tag{6.12}$$

$$\Rightarrow -\lambda x = \ln(1 - R)$$

$$\Rightarrow x = -\frac{1}{\lambda}\ln(1 - R)$$

Now, we generate some random numbers R_i that are uniformly distributed. We have generated 100 uniformly distributed random numbers that are listed in Table 6.1. Here, we are taking R instead of $(1 - R)$ in Equation 6.12, assuming that if R is a uniformly distributed random number, then $(1 - R)$ will also be a uniformly distributed random number.

The respective generated random numbers along with the values of $x = -(1/\lambda)\ln R$ for $\lambda = 1, \lambda = 2, \lambda = 3, \lambda = 4$ and the values of $1 - e^{-\lambda x}$ are all shown in Table 6.1.

The values of Table 6.1 are now plotted on a two-dimensional coordinate system as shown in Figure 6.7.

The plot of $F(x) = 1 - e^{-\lambda x}$ versus x for $\lambda = 1$ (Figure 6.8) and the respective values for a total of 20 sets are given in Table 6.2.

TABLE 6.1
Uniform Random Numbers

R	$x = -\ln R$	$x = -\dfrac{1}{2}\ln R$	$x = -\dfrac{1}{3}\ln R$	$x = -\dfrac{1}{4}\ln R$
0.5557	0.5875	0.2938	0.1958	0.1469
0.7329	0.3107	0.1554	0.1036	0.0777
0.896	0.1098	0.0549	0.0366	0.0275
0.9086	0.0958	0.0479	0.0319	0.024
0.6443	0.4396	0.2198	0.1465	0.1099
0.9585	0.0424	0.0212	0.0141	0.0106
0.897	0.1087	0.0543	0.0362	0.0272
0.4053	0.9031	0.4516	0.301	0.2258
0.8668	0.1429	0.0715	0.0476	0.0357
0.8996	0.1058	0.0529	0.0353	0.0265
0.6052	0.5022	0.2511	0.1674	0.1255
0.6494	0.4317	0.2159	0.1439	0.1079
0.7322	0.3117	0.1558	0.1039	0.0779
0.7577	0.2775	0.1387	0.0925	0.0694
0.4608	0.7748	0.3874	0.2583	0.1937
0.7273	0.3184	0.1592	0.1061	0.0796
0.504	0.6852	0.3426	0.2284	0.1713
0.6923	0.3677	0.1839	0.1226	0.0919
0.3901	0.9414	0.4707	0.3138	0.2353
0.8983	0.1073	0.0546	0.0357	0.0268
0.4065	0.9002	0.4501	0.3001	0.225
0.8191	0.1995	0.0998	0.0665	0.0499
0.3789	0.9705	0.4852	0.3235	0.2426
0.4077	0.8972	0.4486	0.2991	0.2243
0.7037	0.3514	0.1757	0.1171	0.0878
0.3988	0.9193	0.4596	0.3064	0.2298
0.4763	0.7417	0.3708	0.2472	0.1854
0.4688	0.7576	0.3788	0.2525	0.1894
0.4592	0.7783	0.3891	0.2594	0.1946
0.9313	0.0712	0.0356	0.0237	0.0178
0.9002	0.1051	0.0526	0.035	0.0263
0.7544	0.2775	0.1387	0.0925	0.0694
0.6182	0.4809	0.2405	0.1603	0.1202

(Continued)

TABLE 6.1
(Continued) Uniform Random Numbers

R	$x = -\ln R$	$x = -\dfrac{1}{2}\ln R$	$x = -\dfrac{1}{3}\ln R$	$x = -\dfrac{1}{4}\ln R$
0.7924	0.2327	0.1163	0.0776	0.0582
0.46	0.7765	0.3883	0.2588	0.1941
0.5731	0.5567	0.2783	0.1856	0.1392
0.9441	0.0575	0.0288	0.0192	0.0144
0.8718	0.1372	0.0686	0.0457	0.0343
0.3845	0.9558	0.4799	0.3186	0.239
0.8439	0.1697	0.0849	0.0566	0.0424
0.5072	0.6788	0.3394	0.2263	0.1697
0.4966	0.7	0.35	0.2333	0.175
0.4921	0.7091	0.3545	0.2364	0.1773
0.6645	0.4087	0.2044	0.1362	0.1022
0.4738	0.747	0.3735	0.249	0.1867
0.4519	0.7943	0.3971	0.2648	0.1986
0.6219	0.475	0.2375	0.1583	0.1187
0.8991	0.1064	0.0532	0.0355	0.0266
0.8885	0.1182	0.0591	0.0394	0.0296
0.4575	0.782	0.391	0.2607	0.1955
0.4298	0.8444	0.4222	0.2815	0.2111
0.7176	0.3318	0.1659	0.1106	0.083
0.9388	0.0632	0.0316	0.0211	0.0158
0.9495	0.0518	0.0259	0.0173	0.0129
0.9501	0.0512	0.0256	0.0171	0.0128
0.8738	0.1349	0.0674	0.045	0.0337
0.833	0.1827	0.0914	0.0609	0.0457
0.4415	0.8176	0.4088	0.2725	0.2044
0.3784	0.9718	0.4859	0.3239	0.2429
0.8701	0.1391	0.0696	0.0464	0.0348
0.6338	0.456	0.228	0.152	0.114
0.4294	0.8454	0.4227	0.2818	0.2113
0.3915	0.9378	0.4689	0.3126	0.2344
0.89	0.1165	0.0583	0.0388	0.0291
0.6871	0.3753	0.1876	0.1251	0.0938
0.4816	0.7306	0.3653	0.2435	0.1827
0.8868	0.1201	0.0601	0.04	0.03
0.8137	0.2062	0.1031	0.0687	0.0515
0.5908	0.5263	0.2631	0.1754	0.1316
0.8393	0.1752	0.0876	0.0584	0.0438
0.3938	0.9319	0.466	0.3106	0.233
0.8894	0.1172	0.0586	0.0391	0.0293
0.6433	0.4411	0.2206	0.147	0.1103
0.4336	0.8356	0.4178	0.2785	0.2089

(Continued)

TABLE 6.1

(Continued) Uniform Random Numbers

R	$x = -\ln R$	$x = -\dfrac{1}{2}\ln R$	$x = -\dfrac{1}{3}\ln R$	$x = -\dfrac{1}{4}\ln R$
0.595	0.5192	0.2596	0.1731	0.1298
0.8208	0.1975	0.0987	0.0658	0.0494
0.9133	0.0907	0.0453	0.0302	0.0227
0.4164	0.8761	0.4381	0.292	0.219
0.6675	0.4042	0.2021	0.1347	0.101
0.6996	0.3572	0.1786	0.1191	0.0893
0.8027	0.2198	0.1099	0.0733	0.0549
0.6475	0.4346	0.2173	0.1449	0.1087
0.3765	0.9768	0.4884	0.3256	0.2442
0.8782	0.1299	0.0649	0.0433	0.0325
0.7268	0.3191	0.1596	0.1064	0.0798
0.7454	0.2938	0.1469	0.0979	0.0734
0.7252	0.3213	0.1606	0.1071	0.0803
0.9034	0.1016	0.0508	0.0339	0.0254
0.5226	0.6489	0.3245	0.2163	0.1622
0.912	0.0921	0.046	0.0307	0.023
0.3829	0.96	0.48	0.32	0.24
0.6229	0.4734	0.2367	0.1578	0.1183
0.7979	0.2258	0.1129	0.0753	0.0564
0.4578	0.7813	0.3907	0.2604	0.1953
0.4887	0.716	0.358	0.2317	0.179
0.4915	0.7103	0.3551	0.2368	0.1776
0.6989	0.3582	0.1791	0.1194	0.0896
0.8411	0.173	0.0865	0.0577	0.0433
0.7064	0.3476	0.1738	0.1159	0.0869
0.5934	0.5219	0.2609	0.174	0.1305

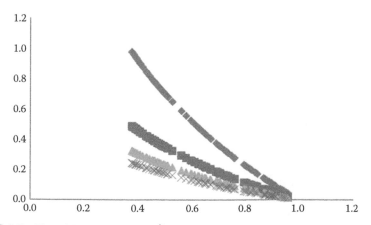

FIGURE 6.7 Plot of R versus $x = -(1/\lambda)\ln(1 - R)$.

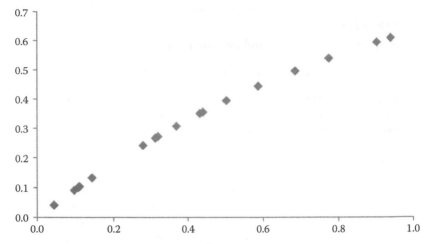

FIGURE 6.8   Plot of x versus $F(x) = 1 - e^{-\lambda x}$.

TABLE 6.2

Values of x and $F(x) = 1 - e^{-\lambda x}$

x	$F(x) = 1 - e^{-\lambda x}$
0.5875	0.4443
0.3107	0.2671
0.1098	0.104
0.0958	0.0914
0.4396	0.3557
0.0424	0.0415
0.1087	0.103
0.9031	0.5947
0.1429	0.1332
0.1058	0.1004
0.5022	0.3948
0.4317	0.3506
0.3117	0.2678
0.2775	0.2423
0.7748	0.5392
0.3184	0.2727
0.6852	0.496
0.3677	0.3077
0.9414	0.6099
0.1073	0.1017

6.2.11.2 Weibull Distribution

The probability density function for the Weibull distribution is given by

$$f(x) = \begin{cases} \dfrac{\beta}{\alpha^\beta} x^{\beta-1} e^{-(x/\alpha)^\beta}, & x \geq 0 \\ 0, & \text{otherwise} \end{cases} \tag{6.13}$$

where:

$\alpha > 0$
$\beta > 0$

We derive the cumulative distribution function by integrating the above probability density function, and we obtain the following cumulative distribution function:

$$F(x) = 1 - e^{-(x/\alpha)^\beta} \text{ for } x \geq 0 \tag{6.14}$$

Next, we set the following equation:

$$F(x) = 1 - e^{-(x/\alpha)^\beta} = R \tag{6.15}$$

Thus, from the above equation, x is expressed in terms of R as follows:

$$F(x) = 1 - e^{-(x/\alpha)^\beta} = R$$

$$\Rightarrow \ln(e^{-(x/\alpha)^\beta}) = \ln(1-R)$$

$$\Rightarrow -\left(\frac{x}{\alpha}\right)^\beta = \ln(1-R) \tag{6.16}$$

$$\Rightarrow \frac{x}{\alpha} = [-\ln(1-R)]^{1/\beta}$$

$$\Rightarrow x = \alpha[-\ln(1-R)]^{1/\beta}$$

Now, with the same set of values of R, x is calculated and the calculated values are given for different sets of α and β in Table 6.3. Unlike the previous example, a total of 20 random numbers have been taken for the Weibull distribution. For the sake of calculation, we are using $x = \alpha(-\ln R)^{1/\beta}$ instead of $x = \alpha[-\ln(1-R)]^{1/\beta}$, assuming that if R is a uniformly distributed random number, then $(1 - R)$ will also be a uniformly distributed random number.

The respective graph is shown in Figure 6.9 and the respective graph between $F(x) = 1 - e^{-(x/\alpha)^\beta}$ and x is similar to the graph shown in Figure 6.8.

TABLE 6.3
Values of Random Number R and x for the Weibull Distribution

R	$x = \alpha(-\ln R)^{1/\beta}$ for $\alpha = 1$ and $\beta = 1$	$x = \alpha(-\ln R)^{1/\beta}$ for $\alpha = 1$ and $\beta = 2$
0.5557	0.5875	0.7665
0.7329	0.3107	0.5574
0.896	0.1098	0.3314
0.9086	0.0958	0.3095

(Continued)

TABLE 6.3

(Continued) Values of Random Number R and x for the Weibull Distribution

R	$x = \alpha(-\ln R)^{1/\beta}$ for $\alpha = 1$ and $\beta = 1$	$x = \alpha(-\ln R)^{1/\beta}$ for $\alpha = 1$ and $\beta = 2$
0.6443	0.4396	0.6630
0.9585	0.0424	0.2059
0.8970	0.1087	0.3297
0.4053	0.9031	0.9503
0.8668	0.1429	0.3780
0.8996	0.1058	0.3253
0.6052	0.5022	0.7087
0.6494	0.4317	0.6570
0.7322	0.3117	0.5583
0.7577	0.2775	0.5268
0.4608	0.7748	0.8802
0.7273	0.3184	0.5643
0.5040	0.6852	0.8278
0.6923	0.3677	0.6064
0.3901	0.9414	0.9703
0.8983	0.1073	0.3276
0.4065	0.9002	0.9488
0.8191	0.1995	0.4467
0.3789	0.9705	0.9851
0.4077	0.8972	0.9472
0.7037	0.3514	0.5928
0.3988	0.9193	0.9588
0.4763	0.7417	0.8612
0.4688	0.7576	0.8704
0.4592	0.7783	0.8822
0.9313	0.0712	0.2668

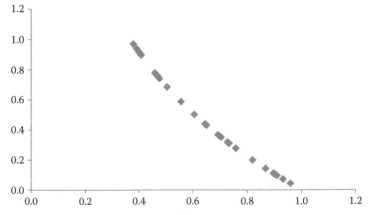

FIGURE 6.9 Plot of R versus $x = \alpha(-\ln R)^{1/\beta}$ for $\alpha = 1$ and $\beta = 1$.

6.2.11.3 Uniform Distribution

The probability density function for the uniform distribution is given by

$$f(x) = \begin{cases} \dfrac{1}{b-a}, & a \leq x \leq b \\ 0, & \text{otherwise} \end{cases} \tag{6.17}$$

The cumulative distribution function for this density function is

$$F(x) = \begin{cases} 0, & x < a \\ \dfrac{x-a}{b-a}, & a \leq x \leq b \\ 1, & x > b \end{cases} \tag{6.18}$$

Next, we set the following equation and find the value of x in terms of R:

$$F(x) = \frac{x-a}{b-a} = R$$
$$\Rightarrow x = a + R(b-a) \tag{6.19}$$

Now using the same set of random numbers as the values of R in Table 6.3, we find the values of x as shown in Table 6.4.

TABLE 6.4

Values of Random Number R and x for the Uniform Distribution

R	$x = a + R(b-a)$ for $a = 0.2$ and $b = 0.6$
0.5557	0.4223
0.7329	0.4932
0.896	0.5584
0.9086	0.5634
0.6443	0.4577
0.9585	0.5834
0.8970	0.5588
0.4053	0.3621
0.8668	0.5467
0.8996	0.5598
0.6052	0.4421
0.6494	0.4598
0.7322	0.4929
0.7577	0.5031
0.4608	0.3843
0.7273	0.4909
0.5040	0.4016

(Continued)

TABLE 6.4

(Continued) Values of Random Number R and x for the Uniform Distribution

R	$x = a + R(b-a)$ for $a = 0.2$ and $b = 0.6$
0.6923	0.4769
0.3901	0.356
0.8983	0.5593
0.4065	0.3626
0.8191	0.5276
0.3789	0.3516
0.4077	0.3631
0.7037	0.4815
0.3988	0.3545
0.4763	0.3905
0.4688	0.3875
0.4592	0.3837
0.9313	0.5725

6.2.11.4   Triangular Distribution

The probability density function for the triangular distribution is given by

$$f(x) = \begin{cases} x, & 0 \le x \le 1 \\ 2-x, & 1 \le x \le 2 \\ 0, & \text{otherwise} \end{cases} \tag{6.20}$$

The cumulative distribution function for the above density function is

$$F(x) = \begin{cases} 0, & x \le 0 \\ \dfrac{x^2}{2}, & 0 < x \le 1 \\ 1 - \dfrac{(2-x)^2}{2}, & 1 < x \le 2 \\ 1, & x > 2 \end{cases} \tag{6.21}$$

Thus, we can have the following:

$$F(x) = \frac{x^2}{2} = R \Rightarrow x = \sqrt{2R} \text{ for } 0 < x \le 1 \tag{6.22}$$

$$F(x) = 1 - \frac{(2-x)^2}{2} = R \Rightarrow x = 2 - \sqrt{2R} \text{ for } 1 < x \le 2 \tag{6.23}$$

As in the previous distributions, we are assuming R instead of $(1 - R)$, assuming that if R is uniformly distributed, then $(1 - R)$ will also be uniformly distributed. Table 6.5 shows the series of triangularly distributed random numbers x generated from a source of uniformly distributed random numbers R.

TABLE 6.5

Values of Random Number R and x for the Triangular Distribution

R	$x = \sqrt{2R}$	$x = 2 - \sqrt{2R}$
0.5557	1.0542	0.9458
0.7329	1.2107	0.7893
0.896	1.3387	0.6613
0.9086	1.348	0.652
0.6443	1.1352	0.8648
0.9585	1.3846	0.6154
0.8970	1.3394	0.6606
0.4053	0.9003	0.1.0997
0.8668	1.3167	0.6833
0.8996	1.3413	0.6587
0.6052	1.1002	0.8998
0.6494	1.1396	0.8604
0.7322	1.2101	0.7899
0.7577	1.231	0.769
0.4608	0.96	1.04
0.7273	1.2061	0.7939
0.5040	1.004	0.996
0.6923	1.1767	0.8233
0.3901	0.8833	1.1167
0.8983	1.3404	0.6596

The inverse transform technique can also be applied to discrete probability distributions. For such a distribution, individual values are summed up instead of integrating them. An example may clarify the concept.

Example 1

Consider the probability mass function given by $f(x) = x/k$ for $x = 1, 2, ..., k$.
The cumulative distribution function can be obtained by

$$F(x) = \sum_{i=1}^{x} f(x)$$

$$= \sum_{i=1}^{x} \frac{i}{k} = \frac{1}{k} \sum_{i=1}^{x} i \qquad (6.24)$$

$$= \frac{1}{k}(1 + 2 + \cdots + x)$$

$$= \frac{1}{k} \frac{x(x+1)}{2} = \frac{x(x+1)}{2k}$$

TABLE 6.6

Values of Random Number R and x for Example 1

R	$\sqrt{1+8kR}$ for $k=2$	$x = \dfrac{-1+\sqrt{1+8kR}}{2}$
0.5557	3.1450	1.0725
0.7329	3.5674	1.2837
0.896	3.9161	1.4581
0.9086	3.9418	1.4709
0.6443	3.3629	1.1814
0.9585	4.0418	1.5209
0.8970	3.9182	1.4591
0.4053	2.7358	0.8679
0.8668	3.8560	1.4280
0.8996	3.9235	1.4617

Next, we can set $F(x) = x(x+1)/2k = R$, from which x can be expressed in terms of R as follows:

$$\frac{x(x+1)}{2k} = R$$

$$\Rightarrow x^2 + x - 2kR = 0$$

$$\Rightarrow x = \frac{-1 \pm \sqrt{1 - 4 \times 1 \times (-2kR)}}{2} \qquad (6.25)$$

$$\Rightarrow x = \frac{-1 \pm \sqrt{1 + 8kR}}{2}$$

The positive root of the above equation may be

$$x = \frac{-1 + \sqrt{1 + 8kR}}{2} \qquad (6.26)$$

From the above equation, it is observed that if $\sqrt{1 + 8kR} > 1$, then we will get a positive root. Table 6.6 gives a set of such values of x based on the values of R.

6.2.12 CONVOLUTION METHOD

In the convolution method [1], the original variable is expressed as a sum of two or more variables. The probability distribution of this sum of variables is known as the convolution of the original variable that belongs to the desired distribution. It has been observed that the binomial variable of a binomial distribution and the Erlang variable of an Erlang distribution can be expressed as the sum of more than one variable. Thus, Sections 6.2.12.1 and 6.2.12.2 show the generation of random variates from these two distributions.

6.2.12.1 Erlang Distribution

The Erlang variable X in the Erlang distribution can be expressed as the sum of several exponentially distributed variables X_i and can thus be represented as

$$X = \sum_{i=1}^{\lambda} -\frac{1}{\lambda\theta} \ln R_i \tag{6.27}$$

Since the Erlang variable has parameters (λ, θ), each X_i will have parameter $(\lambda\theta)$ and can thus be represented as

$$X_i = -\frac{1}{\lambda\theta} \ln R_i \tag{6.28}$$

Replacing the above equation in Equation 6.27, we obtain the following:

$$\begin{aligned} X &= \sum_{i=1}^{\lambda} -\frac{1}{\lambda\theta} \ln R_i \\ &= -\frac{1}{\lambda\theta} \ln\left(\prod_{i=1}^{\lambda} R_i \right) \end{aligned} \tag{6.29}$$

Now, we can generate random numbers R_i and X_i following the Erlang distribution using the above equation.

6.2.12.2 Binomial Distribution

The binomial distribution $\text{Bin}(k, p)$ is equal to the sum of k independent and identically distributed Bernoulli random variables $B(p)$. Thus, we can generate k Bernoulli random variables and sum up those k random variables to obtain the binomial distribution $\text{Bin}(k, p)$. Thus, the steps are as follows:

1. Generate $X_1, X_2, ..., X_k$ Bernoulli variates.
2. Obtain the binomial variates by $Y = X_1 + X_2 + \cdots + X_k = \sum_{i=1}^{k} X_i$

6.2.13 ACCEPTANCE–REJECTION METHOD

The acceptance–rejection method is based on some prespecified condition. If the condition is satisfied, then we accept the generated random number; otherwise we reject the random number. The technique may be described through the following steps:

1. Generate an appropriate condition.
2. Generate a random number R.
3. If R satisfies the specified condition, then accept R, otherwise reject R.

The type of condition depends on the distribution. This method can be applied to Poisson and gamma distributions. Thus, the application of this method on a Poisson distribution is shown in Section 6.2.13.1.

6.2.13.1 Poisson Distribution

The Poisson distribution is given by

$$P(X = x) = \frac{e^{-\lambda}\lambda^x}{x!} \text{ for } x = 0,1,2,... \qquad (6.30)$$

where:
   λ is called the mean and $\lambda > 0$

The cumulative distribution function is given by

$$F(x) = e^{-\lambda}\sum_i \frac{\lambda^i}{i!} \qquad (6.31)$$

The basic steps for generating random variates with a Poisson distribution are as follows:

1. Let $P = 1$ and $i = 0$.
2. Generate a uniformly distributed random number R_i.
3. Set $P = P \times R_i$.
4. If $P < e^{-\lambda}$, then $X = i$; otherwise go to step 5.
5. Increment $i = i + 1$ and go to step 2.

The algorithm can be realized from a simple example. Let us consider $\lambda = 0.3$ and thus $e^{-\lambda} = e^{-0.3} = 0.7408$. For the values of R_i, we will use the previously mentioned random numbers. Thus, we have the following steps:

Step 1: We have $P = 1$ and $i = 0$.
   Generate random number $R_1 = 0.5557$.
   Calculate $P = P \times R_1 = 1 \times 0.5557 = 0.5557$.
   Since $P = 0.5557 < e^{-0.3} = 0.7408$, we accept $X = i = 0$.
   Increment $i = i + 1 = 0 + 1 = 1$.
Step 2: Generate random number $R_2 = 0.7329$.
   Calculate $P = P \times R_2 = 0.5557 \times 0.7329 = 0.4073$.
   Since $P = 0.4073 < e^{-0.3} = 0.7408$, we accept $X = i = 1$.
   Increment $i = i + 1 = 1 + 1 = 2$.
Step 3: Generate random number $R_3 = 0.896$.
   Calculate $P = P \times R_3 = 0.4073 \times 0.896 = 0.3649$.
   Since $P = 0.3649 < e^{-0.3} = 0.7408$, we accept $X = i = 2$.
   Increment $i = i + 1 = 2 + 1 = 3$.
Step 4: Generate random number $R_4 = 0.9086$.
   Calculate $P = P \times R_4 = 0.3649 \times 0.9086 = 0.3315$.
   Since $P = 0.3315 < e^{-0.3} = 0.7408$, we accept $X = i = 3$.
   Increment $i = i + 1 = 3 + 1 = 4$.

Thus, we can generate the random variates for Poisson distribution.

6.3 CONCLUSION

In this chapter, various methods of generating random variates have been introduced. The most popular methods discussed in this chapter are the inverse transform technique, the convolution method, and the acceptance–rejection method. The inverse transform technique actually assigns the cumulative distribution function to a random variable R and expresses the variable X in terms of R. The convolution method is applicable to those variables that can be expressed as the sum of more than one variable. The acceptance–rejection method is based on a certain condition. If the condition is satisfied, then the generated random number or a form of generated random number is accepted; otherwise it is rejected. Some examples for each of these methods have also been cited. The basic purpose is to make the reader familiar with these methods in an easier way.

REFERENCE

1. Law, A. M. and Kelton, W. D. (2003). *Simulation Modeling and Analysis*. 3rd Edition. New Delhi, India: Tata McGraw-Hill.

6.3 CONCLUSION

7 Steady-State Behavior of Stochastic Processes

7.1 INTRODUCTION

The background of this chapter relates to analyzing the output of a simulation study. The random nature of a simulation output is certainly dependent on the nature of the input and the probability distributions used in the study for various parameters. However, sufficient literature is available on a simple output analysis of a simulation study [1]. But the steady-state behavior of various processes related to the simulation output has not been discussed in sufficient detail. Therefore, this chapter discusses various stochastic processes and their steady-state behaviors. The target is to present the readers with a clear understanding of the various stochastic processes and their behaviors in simple terms without much mathematical details so that readers of any background can understand the terms and concepts. Before presenting the steady-state conditions in various disciplines, let us first look at the idea of "steady state."

A system is in steady state if the properties of the system stay unchanged over time. In this state, the system may be in equilibrium. This simply indicates that for any property p, the partial derivative of p with respect to time is zero, as shown in the following equation:

$$\frac{\partial p}{\partial t} = 0 \qquad (7.1)$$

However, the property of a system varies among the various fields of study. If a system is in steady state, then the recently shown behavior of the system will continue to have almost the same value in the future. If the system property is stochastic in nature, then the steady-state behavior will, in general, mean that the probability associated with the property of the system will remain unchanged over time after the system reaches steady state.

Let us consider a simple example of filling up a container with water. When the water is pouring into the container and the container is not yet filled, the system is in transient state, not in the steady-state condition at all, since the volume of water in the container is changing continuously because of pouring water into the container. But once the container is filled completely, then even if the water is pouring continuously into the container, the volume of water will not change, since the container cannot hold any more water. Thus, after the container gets filled, the volume of water in the container reaches a steady-state level, which will remain unchanged now over time, even if the water continues to pour in. This is the clear meaning of the steady-state condition or behavior of any property of a system. Thus, for any field of study, we will have to first identify the variable whose steady-state behavior is being watched. Then, we will have to identify the condition when the value of that variable will not change over time and that value will describe the steady-state behavior of that variable.

The remaining sections of this chapter are arranged as follows: Section 7.2 provides the definition and classification of a stochastic process. Section 7.3 outlines the steady-state conditions and their behaviors in various fields of study. Section 7.4 presents the characteristics of various stochastic processes. Sections 7.5 and 7.6 provide the examples of Markov and Poisson processes. Section 7.7 provides the conclusion for this chapter.

7.2   DEFINITION OF STOCHASTIC PROCESS

Let (Ω, F, P) be a given probability space. A collection of random variables $\{x(t), t \geq o\}$ defined on the probability space is called a stochastic process. A stochastic function is also defined as a function of two arguments $x(w,t)$, $w \in \Omega$, $t \in T$.

A stochastic process is also called a chance process or a random process. Here, the set T is called the parameter space where $t \in T$ denotes the time, the length, the distance, or any other quantity. The set S is the set of all possible values of $x(t)$ for all t and is called the state space where $x(t): \Omega \to A_t$ and $A_t \subseteq R$ and $S = \bigcup_{t \in T} A_t$.

Here, T has finite, countably finite, or uncountably many elements. S also has finite, countably finite, or uncountably many elements, in which case S is a set of intervals on a real line or it could be the whole real line itself.

There are two approaches to a stochastic process, which are as follows:

1. Stochastic process as a family of random variables. This may be represented as $\{x(.,t), t \in T\}$. This approach is an easier approach since we can implement it easily once we come to know the values of t.
2. Stochastic process as a set of functions on T. This is represented as $\{x(w,.), w \in \Omega\}$. This representation is the realization of the process or trajectory or simple path function. The possible outcome is w.

So, we can have two different stochastic processes by fixing either t $[x(.,t)]$ or w $[x(w,.)]$. The stochastic process $\{x(t), t \geq 0\}$ can be one-dimensional, two-dimensional, or n-dimensional. A two-dimensional stochastic process may be represented as follows:

$$x(t) = \{x_1(t), x_2(t)\} \tag{7.2}$$

where:
$x_1(t)$ may represent the maximum humidity of a place
$x_2(t)$ may represent the minimum humidity of the same place

An n-dimensional stochastic process may be represented as follows:

$$x(t) = \{x_1(t), x_2(t), ..., x_n(t)\} \tag{7.3}$$

A stochastic process may also be a complex value stochastic process that contains a complex number component and may be represented, for example, as follows:

$$x(t) = \{x_1(t) + ix_2(t)\} \qquad (7.4)$$

where:
$i = \sqrt{-1}$ is a complex number

Section 7.2.1 briefly describes the various types of stochastic processes.

7.2.1 CLASSIFICATION OF STOCHASTIC PROCESSES

Stochastic processes may be classified based on the value of T, the parameter space [set of $x(t)$], or the value of S, the state space (set of t). Based on the values of T and S, stochastic processes can be divided into the following divisions [2]:

1. Discrete time, discrete space (DTDS) stochastic processes
2. Discrete time, continuous state (DTCS) stochastic processes
3. Continuous time, discrete state (CTDS) stochastic processes
4. Continuous time, continuous state (CTCS) stochastic processes

Each of these types is defined in brief as follows:

1. *DTDS stochastic process.* A stochastic process is a DTDS stochastic process when the parameter space T is discrete in nature. This means that the parameter space T contains countably finite or countably infinite values and the state space contains integer values.

    For example, consider the random variable x_n representing the number of customers in a bank after the nth customer departs from the bank. Here, n is the parameter space representing the customer number and x_n is the state space representing the number of customers present in the bank. Thus, T and S may be represented by the sets given as follows:

$$T = \{\text{set of values of } n\} = \{1,2,3,...\}$$

$$S = \{\text{set of values of } x_n\} = \{0,1,2,...\}$$

    Hence, both T and S are discrete valued sets. Thus, the respective stochastic process $\{x_n, n \geq 1\}$ is a DTDS stochastic process.

2. *DTCS stochastic process.* A stochastic process is a DTCS stochastic process when the parameter space T is discrete in nature. This means that the parameter space T contains countably finite or countably infinite values and the state space contains real values or is continuous in nature.

    For example, consider the random variable x representing the content of a reservoir at nth time point. Here, n is the parameter space representing discrete time points and x is the state space representing the amount of water contained in the reservoir. Thus, x may contain any fractional value within the volume limit of the reservoir. x is continuous in nature and t is discrete in nature and may contain values such as $\{1,2,3,...\}$. Thus, the respective stochastic process $\{x(t), t \geq 0\}$ is a DTCS stochastic process.

3. *CTDS stochastic process.* A stochastic process is a CTDS stochastic process when the parameter space T is continuous in nature. This means that the parameter space T contains fractional values and the state space S contains discrete values.

   For example, consider the random variable x_n representing the number of customers in a bank at any time t. The phrase "any time" may indicate "3 minutes from now," "3.2 minutes from now," and so on. Thus, the parameter set T is represented by

$$T = \{t | t \geq 0\} \tag{7.5}$$

The state space S contains the number of customers at any point in time and the values of S may be given by

$$S = \{\text{set of values of } x_n\} = \{0,1,2,...\}$$

Thus, the stochastic process $\{x(t), t \geq 0\}$ represents the CTDS stochastic process and the values of $x(t)$ are the values in the state space S.

4. *CTCS stochastic process.* A stochastic process is a CTCS stochastic process when the parameter space T is continuous in nature. This means that the parameter space T contains fractional values and the state space S is also continuous in nature, that is, it contains fractional values.

   For example, consider the random variable x representing the content of a reservoir at any time. Here, as earlier, the phrase "any time" may indicate "3 minutes from now," "3.2 minutes from now," and so on. Thus, the parameter set T is represented by

$$T = \{t | t \geq 0\} \tag{7.6}$$

The state space S represents the content of the reservoir at any time and may contain any value between 0 (no water in reservoir) and a maximum value (volume of the reservoir). Thus, if the volume of the reservoir is V, then the state space S may be represented by

$$S = \{\text{set of values representing the volume of water in the reservoir}\}$$

$$= \{x | 0 < x < V\}$$

Before delving into the depth of various stochastic processes, some important definitions related to stochastic processes need to be clarified:

1. *Mean function.* It is defined by the following expression:

$$m(t) = E[x(t)] \tag{7.7}$$

where:

$E[x(t)]$ is basically the expectation of the random variable $x(t)$

2. *Second-order stochastic process.* A stochastic process is a second-order stochastic process if it satisfies the following condition:

$$E[x^2(t)] < \infty \tag{7.8}$$

This means that the second-order moment of $x(t)$ is finite.

3. *Covariance function.* The covariance function of the two variables $x(s)$ and $x(t)$ can be given by

$$c(s,t) = \text{Cov}[x(s), x(t)] \tag{7.9}$$

The above expression can be expressed in terms of expectations, which is shown in the following expression:

$$\begin{aligned} c(s,t) &= \text{Cov}[x(s), x(t)] \\ &= E[x(s)x(t)] - E[x(s)]E[x(t)] \end{aligned} \tag{7.10}$$

The above expression clearly indicates that this is a second-order stochastic process. It satisfies the following properties:

a. $c(s,t) = c(t,s), \forall t, s \in T$.
b. Following Schwartz's inequality, the second property is as follows:
    $c(s,t) \le \sqrt{c(s,s)c(t,t)}$.
c. Covariance is nonnegative definite. This means that if $a_1, a_2, ..., a_n$ is a set of real numbers and $t_i \in T$, we have

$$\sum_{j=1}^{n}\sum_{k=1}^{n} a_j a_k c(t_j, t_k) = E\left[\sum a_j x(t_j)\right]^2 \tag{7.11}$$

Since the expectation on the right-hand side is positive $\{E[\sum a_j x(t_j)]^2 \ge 0\}$, we can conclude that the covariance of a function is a nonnegative definite.

d. The sum and product of any two covariance functions are also covariance functions.

4. *Autocorrelation function.* Autocorrelation between two variables $x(t)$ and $x(s)$ is given by the following expression:

$$R(s,t) = \frac{E[x(t)x(s)] - E[x(t)]E[x(s)]}{\sqrt{\text{Var}[x(t)]}\sqrt{\text{Var}[x(s)]}} \tag{7.12}$$

This function describes the relationship among the values of the process at different time points s and t. If $R(s,t)$ depends only on $|t - s|$, we have

$$R(\tau) = E[x(t) - \mu] \times \frac{[x(t+\tau) - \mu][x(t+\tau) - \mu]}{\sigma^2} \tag{7.13}$$

and we also have the following:

$$m(t) = E[x(t)] = \mu \tag{7.14}$$

and

$$\mathrm{Var}[x(t)] = \sigma^2 \tag{7.15}$$

The relation that should be noted is as follows:

$$R(\tau) = R(-\tau) \tag{7.16}$$

The covariance function is used in time series analysis and signal processing.

5. *Independent increments.* This property says that if, for every $t_1 < t_2 < \cdots < t_n$, $x(t_2) - x(t_1), x(t_3) - x(t_2), \ldots, x(t_n) - x(t_{n-1})$ are mutually independent random variables for all n, then $\{x(t), t \in T\}$ has an independent increment property and the increment is also a random variable.

6. *Ergodic property.* The time average of a function for a sample exists almost everywhere and is related to the space average. This means that whenever a system or a stochastic process is ergodic, the time average is the same for almost all initial points, that is, the process evolves for a longer time and forgets its initial state. This property is called a "memoryless property."

   Statistical sampling can be performed at one instant across a group of identical processes or sampled over time on a single process with no change in the measured result. Ergodic property is important for stationary or Markov property.

7.3   STEADY-STATE CONDITIONS IN VARIOUS FIELDS

This section describes the concept of the term "steady state" for various fields of study. It will also serve as examples of the steady-state behavior of the respective fields of study.

7.3.1   STEADY-STATE CONDITION IN ECONOMICS

The steady-state economy of a country, a state, a region, or a place simply means that the economy is attaining a stable size. This means that there is no significant change in the size of the population of the place and the total consumption does not change over time and within the capacity limit of the economy. So the properties identified here are population size and consumption amount. The values of both of these properties remain unchanged after the economy enters the steady-state condition.

7.3.2   STEADY-STATE CONDITION IN CHEMISTRY

To understand the steady-state condition of a chemical process, first the state variables have to be identified. The steady-state condition will mean that the values of these state variables will remain unchanged over time. To maintain this steady state in the system, there must be a constant flow through the system that keeps the state variables in the steady state. In this case, the conditions of the apparatus and the other related conditions must not change to maintain the steady-state condition. During the period of steady-state condition, there must not be any accumulation of mass or energy. The thermodynamic properties may change from point to point but must stay unchanged at any given point. This is the meaning of a steady-state condition in chemical processes. The example of filling a container with water cited in Section 7.1 can be a good example in this regard.

7.3.3 STEADY-STATE CONDITION IN ELECTRONICS

The steady-state conditions of various properties in an electronic system are very important since many design specifications are set in terms of the steady-state values of those properties. Let us consider the example of the vibration of a particular electronic component. Thus, the property considered in this case is vibration. Vibration reaches a steady-state condition if its value becomes zero, that is, the component becomes motionless and continues to be motionless or the vibration reaches a particular frequency that does not change over time. Steady-state analysis in electronics is very important for its design process. A circuit or network reaches a steady-state condition when the transient conditions are no longer effective on the properties of the system.

7.3.4 STEADY-STATE CONDITION IN ELECTRICAL SYSTEMS

The steady-state condition in electrical systems is basically concerned with the stability of electrical machines or power systems. The related properties of a power system may be voltage, frequency, and phase sequence. Power is generated by a generator, and the generator is synchronized with the same values of voltage frequency and phase sequence. Thus, the steady state of the power system can be described as the ability to return to the original steady state without losing synchronization. The states of a power system can be categorized into steady state, transient state, and dynamic state stability. The steady-state condition is related to small and gradual changes in the system. The steady state is the ability of a system to regain its original stable state.

Now that we have a clearer idea of the meaning of the phrase "steady state" and how it varies across various systems, in the next section we depict the characteristics of various stochastic processes.

7.4   VARIOUS STOCHASTIC PROCESSES

This section presents the characteristics of various stochastic processes. A stochastic process is the study of random variables over time. It may be either discrete or continuous. Discrete stochastic processes are concerned with the study of random variables at discrete points in time. Continuous stochastic processes are concerned with the study of random variables at any point in time. An example of a discrete stochastic process may be the observation of the price of an Intel machine at the beginning of a day. An example of a continuous stochastic process is the study of the number of people arriving to a city center on any given day. The stochastic processes that will be discussed in this chapter are as follows:

- Markov process
- Poisson process
- Gaussian process
- Brownian motion

These processes are discussed since we frequently come across them in our regular study. Thus, Sections 7.4.1 through 7.4.4 discuss these stochastic processes.

7.4.1 MARKOV PROCESS

The study of Markov processes is the study of Markov chains. A Markov chain is a discrete stochastic process if the following condition is satisfied for all states and at all time $t = 0, 1, 2, 3,\ldots$:

$$P(X_{t+1} = i_{t+1} \mid X_t = i_t, X_{t-1} = i_{t-1},\ldots,X_1 = i_1, X_0 = i_0) = P(X_{t+1} = i_{t+1} \mid X_t = i_t) \quad (7.17)$$

The above expression means that the probability distribution of a state at time $t + 1$ depends on the probability distribution of the state at time t and does not depend on any other previous state. In addition, the probability p_{ij} can be defined as the probability of transition from state i in one period to state j in the next period. Furthermore, suppose q_i is the probability that the Markov chain will be at state i at time 0, that is, $P(x_0 = i) = q_i$.

The set of transition probabilities among various states for a Markov chain are shown by a state transition matrix. Thus, a transition matrix may look like the following expression:

$$P = \begin{bmatrix} p_{11} & p_{12} & \cdots & p_{1n} \\ p_{21} & p_{22} & \cdots & p_{2n} \\ \cdots & \cdots & \cdots & \cdots \\ p_{m1} & p_{m2} & \cdots & p_{mn} \end{bmatrix} \quad (7.18)$$

where the sum of probabilities in each row is 1, that is, $\sum_{j=1}^{n} p_{ij} = 1$.

Thus, the probability of being in state j at time t is given by

$$P_j(t) = \sum_{i=1}^{m} q_i p_{ij}(t) \quad (7.19)$$

7.4.2 POISSON PROCESS

A Poisson process is simply a stochastic counting process $\{N(t), t \geq 0\}$, and it may be a discrete or continuous process. It counts discrete events that take place at discrete points in time. Therefore, Poisson values are discrete nonnegative integer values counted at discrete points in time $t \geq 0$. Thus, if we want to know the number of events occurring during the time interval (t_1, t_2), then we will simply calculate $N(t)_2 - N(t)_1$, where $t_2 \geq t_1$.

There is a standard definition for this process. A Poisson process $\{N(t), t \geq 0\}$ is a counting process with the following properties:

1. Initially (at $t = 0$), there is no event, that is, $N(0) = 0$.
2. It has stationary and independent increments.
3. The probability of n number of events occurring at any time t is given by
   $P[N(t) = n] = [(\lambda t)^n / n!] e^{-\lambda t}$.
4. The mean of a Poisson process is λt.
5. The variance of a Poisson process is λt.

The meaning of an "independent increment" is that the number of events occurring at any discrete point in time is independent of the number of events in other discrete points in time, and the number of events keeps increasing with the progress of time. Thus, the only change in the process is the unit jump upward.

7.4.3 GAUSSIAN PROCESS

A Gaussian process basically follows a Gaussian distribution given by the following expression:

$$\frac{1}{2\pi\sigma^2} e^{\frac{-(x-\mu)^2}{2\sigma^2}} \tag{7.20}$$

Like the other modeling techniques, some design decisions are required to set up a Gaussian process. A set of training data is set aside and is used for each test point whenever an inference is drawn with the Gaussian process. As a result, it is evident that the extent of computation can be reduced by the reduction in training data, which in turn may reduce the execution time when implementing through a computer. It is also necessary to choose the dynamic inputs that will be used for the Gaussian process prediction.

7.4.4 BROWNIAN MOTION

A Brownian motion $\{w(t), t \geq o\}$ is a stochastic process, with t as a positive real number, defined on a specified probability space. It has the following properties:

1. At $t = 0$, $w(t) = 0$, that is, $w(0) = 0$.
2. The process $w(0) = 0$ has a stationary independent increment.
3. The increment follows a normal distribution.

The term "independent increment" means that for every nonnegative real number, the incremented random variables are jointly independent. This concept will be made clearer by the processes discussed in Sections 7.4.5 through 7.4.8.

In addition to the above-mentioned four most popular processes, there are numerous other stochastic processes. Some of them include the following:

1. Bernoulli process
2. Simple random walk
3. Population process
4. Stationary process
5. Autoregressive process

Among all the above-mentioned basic types of stochastic processes, the Markov process and the Poisson process are the widely used stochastic processes as evident from the existing literature. Thus, Sections 7.4.5 through 7.4.8 are devoted to a brief overview of processes such as the Bernoulli process, the simple random walk, the population process, the stationary process, and the autoregressive process.

7.4.5 BERNOULLI PROCESS

The Bernoulli process can be created by a sequence of random variables. Let us consider the stochastic process $\{x_i, i = 1,2,...\}$, where x_i is an independent and identically distributed random variable. For the Bernoulli process, each x_i is Bernoulli distributed with parameter p, that is, $x_i \sim B(1,p)$, which clearly means that the Bernoulli process follows a binomial distribution.

If x_i is a Bernoulli random variable, then the sum of the values of x_i is also a Bernoulli variable [$S_n \sim B(n,p)$] and can be represented by

$$S_n = \sum_{i=1}^{n} x_i \tag{7.21}$$

We can create a stochastic process with $\{S_n, n \geq 1,2,...\}$, which is basically a binomial process. Thus, the mean of this distribution is

$$E(S_n) = np \tag{7.22}$$

and the variance is given by

$$\mathrm{Var}(S_n) = np(1-p) \tag{7.23}$$

Thus, we are relating a binomial distribution to the Bernoulli process. Similarly, we can relate a geometric distribution or Pascal's distribution to the Bernoulli process.

7.4.6 SIMPLE RANDOM WALK AND POPULATION PROCESSES

Let (Ω, F, P) be a given probability space. Let us also assume that x_i is a discrete-type independent and identically distributed random variable. Now as a special case, we can define the probability in the following way:

$$P(x_i = m) = \begin{cases} p, & k = 1 \\ (1-p), & k = -1 \end{cases} \qquad (7.24)$$

where:

$P(x_i = m)$ means the probability for $x_i = m$ and $0 < p < 1$

Now, the sum of the variables may be defined as $S_n = \sum_{i=1}^{n} x_i$. Thus, as earlier, we can say that $\{S_n, n \geq 1, 2, \ldots\}$ is also a stochastic process.

Let us consider the population of a particular species of deer in a country. The population size may represent a stochastic process $\{S_n, n \geq 1, 2, \ldots\}$, where S_n represents the total number of deer. The process $\{S_n, n \geq 1, 2, \ldots\}$ is called a population process. Since both n and S_n are discrete in nature, this process is a DTDS stochastic process.

7.4.7 STATIONARY PROCESS

A stationary process is a stochastic process in which probable laws remain unchanged through shifts in time and space. A stationary process is especially useful for time series data.

Time series is a collection of random variables indexed by time. It is a special case of a stochastic process $\{x(t), t \geq 0\}$. One of the main characteristics of time series is the interdependency of observations over time. This interdependency needs to be accounted for in the time series data modeling to improve the temporal behavior and in the forecast of future movement.

Using the stationary process, raw data are transformed into stationary data so as to facilitate forecasting using the stationarity property. There are different forms of stationarity depending on the particular statistical properties of the time series. The widely used and applied stationarity are strict-sense and white-sense stationarity property, which are discussed in brief as follows:

1. *Strict-sense stationarity property.* This is the most important stationarity property. This property can be defined as follows:

    If, for arbitrary $t_1, t_2, \ldots, t_n$, the joint distribution of the random vector, that is, $[x(t_1), x(t_2), x(t_3), \ldots, x(t_n)]$ and $[x(t_1 + h), x(t_2 + h), x(t_3 + h), \ldots, x(t_n + h)]$, is the same for all $h > 0$, then we say that the stochastic process $\{x(t), t \geq 0\}$ is a strict-sense stochastic stationarity of order n.

    If the above-mentioned property is satisfied for all n, then we can conclude that the stochastic process is going to be a strict-sense stationary process $\{x(t), t \geq 0\}$ for any integer n.

2. *White-sense/weakly/covariance stationarity property.* A stochastic process is said to be a white-sense or weakly stationary process if it satisfies the following properties:
   a. The expectation $E[x(t)] = m(t)$ or mean is independent of t.
   b. The second-order moment is finite, that is,

$$E[x^2(t)] < \infty \qquad (7.25)$$

   c. The covariance $c(s,t)$ is a function on only the time difference $|t - s|$ for all t and s. This can be expressed mathematically as

$$c(s,t) = f(|t-s|) \qquad (7.26)$$

It should be noted that the white-sense stationary process does not imply the strict-sense stationary process and the strict-sense stationarity property does not imply the white-sense stationarity property.

To make the understanding effective, we give a simple example. Let x_i be an independent and identically distributed random variable and also assume that each x_i is a Bernoulli distributed random variable $B(1,p)$ with parameter p, which is equivalent to saying that the random variable is binomially distributed with parameters 1 and p. We can create a stochastic process $\{x_i, i = 1,2,...\}$ with random variable x_i. Now, we can verify the type of stochastic process $\{x_i, i = 1,2,...\}$. First, we check whether the stochastic process $\{x_i, i = 1,2,...\}$ satisfies the conditions or properties of white-sense stationarity. Thus, we have the following:

1. The mean function for each x_i, $m(i) = E(x)_i = p$ (since it is Bernoulli distributed). Thus, the mean is independent of i.
2. The second-order moment of x_i is given by

$$E(x_i^2) = 1^2.p + 0^2.(1-p) = p \qquad (7.27)$$

   Thus, the second-order moment is finite.
3. Covariance between two random variables x_i and x_j is given by

$$c(i,j) = E(x_i x_j) - E(x_i)E(x_j)$$

$$= E(x_i)E(x_j) - E(x_i)E(x_j) = 0, \forall i \neq j$$

If $i = j$, then we have

$$c(i,j) = E(x_i^2) - (E(x_i))^2 = p(1-p) \text{ (Bernoulli formula)}$$

Thus, it is observed that all three properties for white-sense stationarity have been satisfied. The respective stationary process $\{x_i, i = 1,2,...\}$ is a white-sense stationary process. Now, we check whether the stationary process $\{x_i, i = 1, 2,...\}$ is a strict-sense stationary process. For that purpose, we will have to verify the joint distribution of the variables x_i, that is, the variables $(x_{i_1}, x_{i_2}, x_{i_3}, ..., x_{i_n})$, where x_i follows the Bernoulli distribution $[x_i \sim B(1,p)]$.

Thus, the joint distribution is going to be the product of the individual distributions. Now, if we shift i_1 by an amount h, i_2 by an amount h, and so on, we will get the following random variables:

$$(x_{i_1+h}, x_{i_2+h}, x_{i_3+h}, \ldots, x_{i_n+h})$$

If we find the joint distribution, then, since each variable is independent, the joint distribution is also going to be the product of n random variables, and thus, the distribution is again going to be identical and mutually independent.

Hence, it is proved that the condition for strict-sense stationarity is also satisfied for all $h > 0$ and for all n. Thus, $\{x_i, i = 1, 2, \ldots\}$ is a strict-sense stationary process. This particular stochastic process is a strict-sense stationary process as well as a white-sense stationary process.

7.4.8 AUTOREGRESSIVE PROCESS

Before discussing the autoregressive process, let us take a look at some of the related concepts.

A pure random process or white noise process $\{x_t, t \in T\}$ has the following properties:

1. The mean or expectation is a constant, that is, $E(x_t) = m$.

2. $E(x_t x_{t+k}) = \begin{cases} \sigma^2, & k = 0 \\ 0, & k \neq 0 \end{cases}$

Its covariance is a stationary process.

The concept of a pure random process is utilized in the definition of a moving average process discussed next.

A moving average process is represented as

$$x_t = a_0 e_t + a_1 e_{t-1} + \ldots + a_p e_{t-p} \tag{7.28}$$

where:
$a_0, a_1, \ldots, a_p$ are the real constants
$\{e_t\}$ is a pure random process with mean 0 and variance σ^2

If $a_p \neq 0$, then the stochastic process $\{x_t, t \in T\}$ is called a moving average process of order h.

The covariance function for this process is given by

$$C_p = E(x_t x_{t+p}) \tag{7.29}$$

The above expression can be expanded as follows:

$$C_p = \begin{cases} (a_0 a_p + a_1 a_{p+1} + \cdots + a_{h-p} a_h), & p \leq h \\ 0, & p > h \end{cases}$$

The correlation function is given by

$$\rho_k = \frac{C_p}{C_0}$$

$$= \begin{cases} \dfrac{(a_0 a_p + a_1 a_{p+1} + \cdots + a_{h-p} a_h)}{a_0^2 + a_1^2 + \cdots + a_h^2}, & p \leq h \\ 0, & p > h \end{cases} \tag{7.30}$$

Next, the autoregressive process is given by the following expression:

$$x_t + b_1 x_{t-1} + b_2 x_{t-2} + \ldots + b_n x_{t-n} = e_t \text{ with } b_n \neq 0$$

where:

$x_t = \sum_{i=0}^{\infty} b_i e_{t-i}$ is an autoregressive process of an infinite order

Thus, the autoregressive process of order 1 is given by

$$x_t + b_1 x_{t-1} = e_t \tag{7.31}$$

and the autoregressive process of order 2 is given by

$$x_t + b_1 x_{t-1} + b_2 x_{t-2} = e_t \tag{7.32}$$

This is called the Yule process.

An autoregressive moving average (ARMA) process is a combination of the autoregressive process and the moving average process. Thus, the ARMA process is represented by the following expression:

$$\sum_{i=0}^{p} b_i x_{t-i} = \sum_{j=0}^{q} a_j e_{t-j} \tag{7.33}$$

where:

$b_0 = 1$

The above expression is called the ARMA process of order p and q or, in short, ARMA(p,q). The covariance function for this process is given by

$$C_p = E(x_t x_{t+p}) - E(x_t) E(x_{t+p}) \tag{7.34}$$

and the correlation function for this process is given by

$$\rho_k = \frac{C_p}{\sqrt{\text{Var}(x_t)} \sqrt{\text{Var}(x_{t+p})}} \tag{7.35}$$

7.4.9 EXAMPLES OF THE MARKOV PROCESS

Section 7.4.1 presented a brief introduction to the concept of a Markov process. This section discusses the various aspects of a Markov process.

Consider a coin tossing event. Suppose the probability that a "head" event will occur is p, and thus the probability that a "tail" event will occur is $(1 - p)$. If the coin is tossed n number of times, then for the nth trial, we can say that the probability that the nth event will be a head is $P(x_n = 1) = p$ and naturally the probability that the nth event will be a tail is $P(x_n = 0) = 1 - p$.

Thus, for n trials, we have a sequence of n random variables: $x_1, x_2, \ldots$ This sequence of random variables can make a stochastic process. Here, all x_i are mutually independent. The partial sum of the first n trials can be given by

$$S_n = x_1 + x_2 + \cdots + x_n \tag{7.36}$$

Thus, the sum of the first $(n + 1)$th terms can be given by

$$S_{n+1} = S_n + x_{n+1} \tag{7.37}$$

The sum S_n can form a stochastic process represented by $\{S_n, n = 1,2,\ldots\}$. The conditional probability that a head (probability: p) appears in the $(n + 1)$th trial is given by

$$P(S_{n+1} = k + 1 | S_n = k) = p \tag{7.38}$$

This expression also indicates that

$$P(S_{n+1} = k | S_n = k) = 1 - p, \text{ where } n \geq 0 \tag{7.39}$$

So we can come to the conclusion that

$$P(S_{n+1} = k | S_1 = i_1, S_2 = i_2, \ldots, S_n = i_k) = p(S_{n+1} = k | S_n = k) = 1 - p \tag{7.40}$$

Thus, the occurrence of a particular event at the $(n + 1)$th trial is dependent only on the occurrence of the last event and not on the occurrence of the previous events. This property is commonly known as the memoryless property or the Markov property.

Thus, in the Markov process, the occurrence of any event depends on the occurrence of the last event, and all the previous events' occurrences move out of the memory. Therefore, a stochastic process $\{S_n, n = 1,2,\ldots\}$ satisfying the Markov property is known as a Markov process. If the state space of a Markov chain is discrete in nature, then the Markov chain is called a discrete-type Markov chain (DTMC) or if the state space is of continuous type in nature, then the Markov chain is called a continuous type Markov chain (CTMC).

In a Markov chain, a transition from one state to another state takes place. The probability of being in state j at the nth stage can be represented by $p_j(n) = P(x_n = j)$, which is the probability mass function for the random variable x_n. Now, the conditional probability of transition from state j to state k at the nth stage can be given by

$$P_{jk}(m,n) = P(x_n = k | x_m = j), 0 \leq m \leq n \text{ and } j,k \in S \tag{7.41}$$

When a DTMC is time homogeneous, that is, it satisfies the time-invariant property, then $P_{jk}(m,n)$ depends on the difference $n - m$. Thus, the transition to state k in n steps can be given by

$$P_{jk}(n) = P(x_{m+n} = k \mid x_m = j) \tag{7.42}$$

The above expression is known as an n-step transition probability function. Using this, a one-step transition probability can be defined and is given by

$$P_{jk}(1) = P(x_{m+1} = k \mid x_m = j), n \geq 0 \tag{7.43}$$

and

$$P_{jk}(0) = \begin{cases} 1, & j = k \\ 0, & \text{otherwise} \end{cases} \tag{7.44}$$

Thus, for a total of n states, we can express all the probability values in terms of a matrix. Each cell of the matrix can represent a transition probability from state i to state j and the matrix can be expressed as

$$P = [P_{ij}] \tag{7.45}$$

The matrix represented by the above expression will look like the following:

$$P = \begin{bmatrix} P_{11} & P_{12} & \cdots & P_{1n} \\ P_{21} & P_{22} & \cdots & P_{2n} \\ \cdots & \cdots & \cdots & \cdots \\ P_{n1} & P_{n2} & \cdots & P_{nn} \end{bmatrix} \tag{7.46}$$

where each P_{ij} is expressed as

$$P_{ij} = P(x_{n+1} = j \mid x_n = i), n \geq 0, i, j \in S \tag{7.47}$$

From expressions (7.45) and (7.46), it can easily be observed that the matrix P is a square matrix. It satisfies two properties given as follows:

- $P_{ij} \geq 0, \forall i, j \in S$, which is the basic property of all probability values.
- $\sum_{j \in S} P_{ij} = 1, i \in S$, which means that the row sum for each row of the matrix is 1.

The matrix P is thus a stochastic matrix and this is a pictorial way of viewing a transition matrix. The matrix is known as a transition probability matrix for a DTMC. A state transition matrix is useful for visually studying the properties of DTMC. It can also be shown by a directed graph that is commonly known as a state transition diagram or a directed graph for a DTMC. Two examples that show the transition probability matrices, which are also expressed in terms of state transition diagrams, are as follows:

Example 1

Consider the following transition probability matrix:

$$
\begin{bmatrix}
0 & 2/3 & 1/3 & 0 \\
0 & 0 & 0 & 1 \\
0 & 0 & 0 & 1 \\
1/2 & 0 & 0 & 1/2
\end{bmatrix}
\tag{7.48}
$$

This matrix has four rows and four columns. Thus, there are a total of four states in this matrix. The cells represent the probability values of transition from one state to another state. For example, the probability of transition from state 1 to state 1 is 0. Similarly, the probability of transition from state 1 to state 1, state 1 to state 4, state 2 to state 1, state 2 to state 2, state 2 to state 3, state 3 to state 1, state 3 to state 2, state 3 to state 3, state 4 to state 2, and state 4 to state 3 are all 0. The transition probability from state 1 to state 2 is 2/3; from state 1 to state 3, it is 1/3; from state 2 to state 4, it is 1; for state 3 to state 4, it is 1; from state 4 to state 1, it is 1/2; and from state 4 to state 4, it is 1/2. The corresponding state transition diagram is shown in Figure 7.1.

Example 2

Consider the following transition probability matrix:

$$
\begin{bmatrix}
0 & 0 & 1 & 0 \\
1/2 & 0 & 1/2 & 0 \\
0 & 1/3 & 2/3 & 2/3 \\
0 & 3/4 & 1/2 & 1
\end{bmatrix}
\tag{7.49}
$$

The corresponding state transition diagram is provided in Figure 7.2.

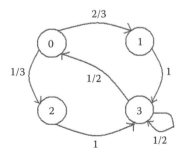

FIGURE 7.1 State transition diagram for Example 1.

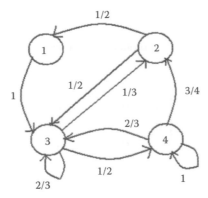

FIGURE 7.2   State transition diagram for Example 2.

The distribution of the probability given by $p_j(n) = P(x_n = j), n = 1,2,\ldots$, and $j \in S$ can be written as

$$p_j(n) = \sum_{i \in S} P(x_0 = i)P(x_n = j \mid x_0 = i) \tag{7.50}$$

The above expression gives the conditional probability of being in state j at the nth step, which is expressed by a marginal distribution $P(x_0 = i)$ and a conditional probability $P(x_n = j \mid x_0 = i)$. Now let the initial probability vector be given by the following expression:

$$P(0) = [P(x_0 = 0)P(x_0 = 1)P(x_0 = 2)\ldots] \tag{7.51}$$

Now the probability of transition from state i to state j in n steps can be expressed as the probability that the random variable x_n will have the value j, provided the initial state of x_n is i $(x_0 = i)$, that is,

$$P_{ij}^{(n)} = P(x_n = j \mid x_0 = i) \tag{7.52}$$

The above expression can be calculated by the Chapman–Kolmogorov equation. Consider the following expression, which was explained earlier. Expression (7.44) gives the probability of state j after $m + n$ steps based on the state i after m steps.

$$P_{ij}^{(n)} = P(x_{m+n} = j \mid x_n = 1) \tag{7.53}$$

From the above expression, we can get the expression for a two-step transition probability as

$$P_{ij}^{(2)} = P(x_{n+2} = j \mid x_n = i) \tag{7.54}$$

This gives the probability of moving from state i to state j in two steps. The above expression can be written as follows:

$$P_{ij}^{(2)} = P(x_{n+2} = j | x_n = i)$$

$$= \sum_{k \in S} P(x_{n+2} = j, x_{n+1} = k | x_n = i)$$

$$= \sum_{k \in S} \frac{P(x_{n+2} = j, x_{n+1} = k, x_n = i)}{P(x_n = i)} \qquad (7.55)$$

$$= \sum_{k \in S} \frac{P(x_{n+2} = j | x_{n+1} = k, x_n = i) P(x_{n+1} = k, x_n = i)}{P(x_n = i)}$$

$$= \sum_{k \in S} \frac{P(x_{n+2} = j | x_{n+1} = k) P(x_{n+1} = k | x_n = i) P(x_n = i)}{P(x_n = i)}$$

In other words, the above expression is same as the following:

$$P_{ij}^{(2)} = \sum_{k \in S} P_{ik} P_{kj}^{(m)} \qquad (7.56)$$

Similarly, by the method of induction, we can write

$$P_{ij}^{(m+1)} = \sum_{k \in S} P_{ik} P_{kj}^{(m)} \text{ or } P_{ij}^{(m+1)} = \sum_{k \in S} P_{ik}^{(m)} P_{kj} \qquad (7.57)$$

The above expression can be generalized through the following expression:

$$P_{ij}^{(m+n)} = \sum_{k \in S} P_{ik}^{(m)} P_{kj}^{(n)} \qquad (7.58)$$

The above expression is known as the Chapman–Kolmogorov equation for a time homogeneous DTMC for transition probability. The matrix for the mth transition probability, as discussed before, is $P = [P_{ij}]$. Thus, the two-step transition probability may be given by $P^{(2)} = p \cdot p = p^2$. Thus, in general, the n-step transition probability may be expressed as

$$P^{(n)} = p^n \qquad (7.59)$$

Thus, we can also write the following:

$$P(n) = [P(x_n = 0) P(x_n = 1) P(x_n = 2)...] \qquad (7.60)$$

The above expression can be generalized as

$$P(n) = P(0) \cdot P^{(n)} = p(0) p^n \qquad (7.61)$$

Thus, we need the initial probability and the n-step transition matrix to give the n-step transition probability and distribution of x_n.

There are several concepts related to the state transitions for a Markov chain. These concepts are given in brief as follows:

1. *Accessibility.* State i is accessible from state j if $P_{ij}^{(n)} > 0$ for some $n \geq 0$. Thus, the probability of ever entering state i, provided that initially the system is in state j, is given by $P\{\bigcup_{n=0}^{\infty} x_n = i | x_0 = j\}$.

2. *Communicate.* Two states are said to communicate with each other, which also means that state i is accessible from state j and state j is accessible from state i, that is, $i \leftrightarrow j$. It satisfies the following properties:
   a. $P_{ii}^{(0)} = P(x_0 = i | x_0 = i) = 1, \forall i$, which can be expressed as $i \leftrightarrow i$. This property is known as the reflexivity property.
   b. If $i \leftrightarrow j$, then $j \leftrightarrow i$. This property is known as the symmetric property.
   c. If $i \leftrightarrow j$ and $j \leftrightarrow k$, then $i \leftrightarrow k$. This property is known as the transitive property.

   The "communicate" property is actually the "equivalence property," and this property divides the set of states into communicating classes.

3. *Class.* A class of states is a subset of the state space S such that every state of the class communicates with every other state and there is no other state outside the class that communicates with all other states in the class. One of the main properties of a class is that all states belonging to a particular class share the same properties.

4. *Periodicity.* State i is a return state if $P_{ii}^{(n)} > 0$ for some $n \geq 1$. The period d_i of a return state i is defined as the greatest common divisor of all m such that $P_{ii}^{(m)} > 0$:

$$d_i = \gcd\{m : P_{ii}^{(m)} > 0\} \tag{7.62}$$

5. *Closed set of states.* If C is the set of states such that no state outside C can be reached from any state in C, then we can say that the set C is closed. If there is only one element in C, then state i is called an absorbing state ($P_{ii}^{(1)} = 1$).

6. *Irreducible.* This concept is valid for both discrete and continuous time Markov chains. This concept says the following:
   a. If a Markov chain does not contain any other closed subset of the state space S, other than the state space S itself, then the Markov chain is called an irreducible Markov chain.
   b. The states of a closed communicating class share the same class properties. Hence, all the states in the irreducible chain are of the same type.

7. *First visit.* This is defined as being in state k for the first time at the nth time step, provided the system starts with state j initially. This can be represented by $f_{jk}^{(n)}$. Thus, we have the following:

$$f_{jk}^{(n)} = P(\text{state } k \text{ for the first time at the } n\text{th time step} \mid \text{the system}$$
$$\text{starts with state } j \text{ initially})$$

and

$$P_{jk}^{(n)} = P(\text{state } k \text{ at the } n\text{th time step}|\text{state } j \text{ initially})$$

Thus, mathematically, we can write:

$$P_{jk}^{(n)} = \sum_{i=0}^{n} f_{jk}^{(i)} P_{kk}^{(n-i)}, \quad n \geq 1 \tag{7.63}$$

where:
$$P_{kk}^{(0)} = 1$$
$$f_{jk}^{(0)} = 0$$
$$f_{jk}^{(1)} = P_{jk}$$

8. *First passage time.* This is defined by the following expression:

$$F_{jk} = P(\text{the system starts with state } j \text{ will ever reach state } k)$$

$$= \sum_{i=1}^{k} f_{jk}^{(n)} \tag{7.64}$$

We have the following two possibilities for F_{jk}:

$$F_{jk} < 1 \text{ or } F_{jk} = 1$$

9. *Mean recurrence time.* It is given by the following expression:

$$\mu_{jj} = \sum_{n=1}^{\infty} n f_{jk}^{(n)} \text{ when } k = j \tag{7.65}$$

If $f_{jj}^{(n)}$ is the distribution of the recurrence time of state j and if $F_{jj} = 1$, then the return to state j is certain. The mean recurrence time for state j is μ_{jj}, where $\mu_{jj} = \sum_{n=1}^{\infty} n f_{jj}^{(n)}$.

In general, there are two types of states: recurrent states and transient states. Definitions of each of these types are given as follows:

a. *Recurrent state*—A state is said to be a recurrent state or a persistent state if

$$F_{jj} = 1 \text{ and } F_{jj} = \sum_{n=1}^{\infty} f_{jj}^{(n)}$$

A recurrent state can be further subdivided into two types based on the first passage time distribution: null recurrent states and positive recurrent states.

If $\mu_{jj} < \infty$, that is, μ_{jj} is of finite value, then the state is called a positive recurrent state, whereas if $\mu_{jj} = \infty$, that is, μ_{jj} is of infinite value, then the state is called a null recurrent state. A positive recurrent state can also be divided into two types: periodic states (if $d_j \neq 1$) and aperiodic states (if $d_j = 1$).

b. *Transient state*—If $F_{jj} < 1$, that is, the probability of returning to state j from state j is not certain, such a state is called a transient state.

7.4.10 AN EXAMPLE OF THE POISSON PROCESS

Let us consider the process of arrival of customers at a barber shop. Let $N(t)$ denote the number of arrivals occurring during the interval $[0,t]$. Thus, the stochastic process $\{N(t), t \geq 0\}$ is a countably infinite CTDS stochastic process. To create a Poisson process, a few assumptions are required.

Let us assume the following:

- Within a small interval of time $(t, t + \Delta t)$, the probability of one arrival is $\lambda \Delta t + o(\Delta t)$ for $\lambda > 0$.
- The probability of more than one arrival is going to be of the order of Δt, that is, $o(\Delta t)$. Thus, as $\Delta t \to 0$, the order $o(\Delta t) \to 0$.
- Nonoverlapping intervals are mutually independent.

Now, we are ready to develop a Poisson process $[N(t), t \geq 0]$. We first divide the interval $[0,t]$ into n equal parts, each of whose length is t/n. Here, we can apply a binomial distribution. Thus, the probability that k arrivals take place in interval $[0,t]$ is given by

$$P[N(t) = k] = \binom{n}{k}\left(\lambda\frac{t}{n}\right)^k \left(1 - \lambda\frac{t}{n}\right)^{n-k} \tag{7.66}$$

As $n \to \infty$, the above expression becomes

$$P[N(t) = k] = \lim_{n \to \infty}\binom{n}{k}p^k(1-p)^{n-k}, \text{ where } p = \lambda\frac{t}{n}$$

$$= \lim_{n \to \infty}\frac{n!}{k!(n-k)!}\left(\frac{\lambda t}{n}\right)^k\left(1 - \frac{\lambda t}{n}\right)^{n-k} \tag{7.67}$$

$$= \lim_{n \to \infty}\frac{n!}{k!(n-k)!}\frac{(\lambda t)^k}{k!}\left(1 - \frac{\lambda t}{n}\right)^n\left(1 - \frac{\lambda t}{n}\right)^{-k}$$

$$= \frac{(\lambda t)^k}{k!}e^{-\lambda t}$$

For fixed t, $N(t) \sim$ Poisson distribution with parameter λt. Each random variable $N(t)$ follows a Poisson distribution so that the collection will be a Poisson process. There are two types of Poisson processes: homogeneous Poisson processes, where λ is a constant, and nonhomogeneous Poisson processes, where λ is not a constant.

7.5 CONCLUSION

In this chapter, a brief overview of various stochastic processes has been provided. The most popular of them are the Markov process and the Poisson process, and thus the concepts for these two stochastic processes have been elaborated much more than the other stochastic processes such as the Bernoulli process, the simple random walk, and the autoregressive process. A brief overview of the stationary process has also been provided. At the beginning of the chapter, a brief overview of the steady-state conditions for various fields of study has been provided without mathematical details for readers' ease of understanding. The introduction of various stochastic processes has gone into some depth of mathematical study that needs the basic understanding of probability theory, basic calculus, combinatorial mathematics, and basic matrix algebra. Thus, this chapter is expected to be beneficial to readers of various levels.

REFERENCES

1. Law, A. M. and Kelton, W. D. (2003). *Simulation Modeling and Analysis*. 3rd Edition. New Delhi, India: Tata McGraw-Hill.
2. Medhi, J. (2009). *Stochastic Processes*. 3rd Edition. New Delhi, India: New Age International Publishers.

8 Statistical Analysis of Steady-State Parameters

8.1 INTRODUCTION

After a simulation study is performed, the results are analyzed. There are always variances in simulation results, and thus care must be taken while interpreting the results of a simulation study. Suppose $Y_1, Y_2, Y_3, \ldots$ are the results of a simulation study. These results may not be independent in general. Moreover, these results may have various distributions based on various factors in the simulation study. Various statistical methods are available for analyzing the output of a simulation study. The particular methods to apply depend on the type of simulation study as well as the context in which the study is conducted. The application of the methods also depends on the simulation software being used. The last point is of vital importance since the present nonphysical simulation studies are all based on computerized simulation software. Thus, the capability and flexibility of such software are essential criteria for the success of simulation studies [1].

However, there are some common statistical measures for analyzing the performance of a system through a simulation study. These measures include various statistical tests, such as the chi-squared test and the F-test, and the calculation of confidence intervals, besides other common measures such as mean, standard deviation, and variance.

While conducting a simulation study, some points must be noted. The first among these is that the results of a simulation study cannot be realized by a single run. Thus, multiple runs are conducted to obtain results that are close to real-world scenarios. The number of runs that are required for obtaining good results is a critical issue in any simulation study. In addition to the number of runs to be conducted, there are other critical parameters. One of them is the setting of the initial parameters and conditions. Simulation is such a study that is context dependent. Thus, such parameters are of vital importance for the simulation to be effective.

Another important issue is deciding the probability distribution for the parameters in a simulation study. There are numerous probability distributions and some of them have been discussed in Chapter 4. But the choice of the probability distributions shapes the output of the simulation study. Thus, emphasis must be placed on choosing the appropriate parameters with appropriate probability distributions.

This chapter presents a brief overview of the types of simulation studies and the respective statistical analyses of the outputs of simulation studies. There are basically two types of simulation studies—terminating simulation and steady-state simulation. Section 8.2 gives a brief overview of these two types of simulation studies.

For ease of reading and understanding, this chapter avoids using mathematical jargon and in-depth mathematical understanding of the concepts. Instead, the chapter's focus is on explaining the main problems faced while performing simulation studies, since most of today's simulation studies are done with some kind of software. Thus, the real problems while designing an analysis of the output are required to be known.

8.2 TERMINATING AND STEADY-STATE SIMULATION

Before conducting the simulation experiment, the user must decide what type of simulation study will have to be performed. The choice depends on the type of problem being studied. The types of simulation studies with respect to the output analysis are shown in Figure 8.1.

In a terminating simulation, the parameters of the simulation study are all well defined. Thus, the specific initial and stopping conditions are also defined beforehand, as these are parts of the simulation study. There are various realistic ways to model these conditions. The output of a terminating simulation depends on the defined initial conditions. Thus, defining the initial parameters and conditions is of vital importance in this type of simulation system.

There are a plenty of real-world examples of terminating simulations. These examples are natural events that also define their own duration, which in turn defines the length of each simulation run. A few such examples are as follows:

- Office hours of an organization: For example, if the office opens at 9:00 a.m. and closes at 6:00 p.m., the duration of office hours is 9 hours. Thus, each simulation event may be defined as 9 hours of simulation: $E = \{9 \text{ hours of simulation}\}$
- A manufacturing organization manufactures 200 parts per day. The total duration of the manufacturing floor will be the time required to manufacture 200 parts in a day. Thus, the simulation event may be defined as the completion time of the 200th part.
- A shopping complex opens at 9:00 a.m. and closes at 11:00 p.m. The total hours of activity of the shopping complex is 14 hours. Thus, the simulation event may be defined as "14 hours of simulation."

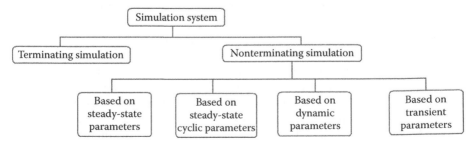

FIGURE 8.1   Types of simulation studies based on output analysis.

Nonterminating simulation does not terminate, as the literal words convey. But there is no realistic event that will not terminate. This means that, for these types of models, we are more interested in the long-run behavior of the model. If the performance measure of the model is the steady-state characteristics and parameters of the model, this type of simulation study can be performed. A perfect nonterminating simulation does not depend on the initial parameters and conditions of the model. But such conditions are only possible in theory. Practically, we can never start a simulation without initializing the parameters of the simulation study. Thus, to give the essence of a real nonterminating simulation, the simulation study is performed for a very long time in general so as to get rid of the effect of the defined initial parameters. The results are taken at sufficiently long times after the simulation actually starts to ensure that the results do not get shaped by the defined initial parameters.

Nonterminating simulations can be categorized into four categories: (1) nonterminating simulation based on steady-state parameters, (2) nonterminating simulation based on steady-state cyclic parameters, (3) nonterminating simulation based on dynamic parameters, and (4) nonterminating simulation based on transient parameters.

A nonterminating simulation based on steady-state parameters is a normal nonterminating simulation in which the parameters are first defined by the simulator and the simulation is first run for a long time to eradicate the effects of the initialization, and after being sufficiently sure that there is no effect of initialization left anymore, the results are taken. For example, consider a 24-hour medical store. Customers arrive at the medical store throughout the entire 24 hours, although the frequency and rate of arrivals vary with time. The performance measure of such a system may be the average daily arrival of customers.

A nonterminating simulation based on steady-state cyclic parameters is a simulation in which the important parameters are cyclic in nature. In simpler terms, it is the type of simulation study in which the change of the relevant parameter values is cyclic in nature. For example, consider the inventory replenishment system of items in a retail store. The replenishment of items in inventory is cyclic in nature. For example, if weekly orders are placed in a retail store, the placement of orders and the arrival of orders form a cycle. By the time the orders arrive, the items are also sold from the inventory and the next order is placed when the stock is decreased to a particular predefined level. The performance of such an inventory system can be measured by the average inventory level.

A nonterminating simulation based on dynamic parameters is a simulation in which the parameters are dynamic in nature. This means that the values of such parameters vary randomly. For example, an airport terminal is always open and the number of passengers varies with the time of day. The total number of passengers changes each hour, week, and month. Moreover, the number of passengers at a particular hour is perfectly dynamic in nature. Thus, the waiting time of passengers is also dynamic in nature, and the performance of such a system may be measured by the average waiting time of the passengers before boarding an airplane.

A nonterminating simulation based on transient parameters is a simulation in which the values of the parameters depend on any transition. For example, the traffic in a day varies with time. The roads may have one-way or two-way traffic. Or in some other cases, one side of a road may be blocked for some reason, such as an accident or the arrival of a VIP guest or for any other reason. Such incidents also affect the traffic on road. Cars may have to wait for a significant amount of time. The performance of such systems may be measured by the average waiting time during a traffic jam on such roads.

From the above discussion and definitions, it is evident that there are distinct differences between terminating and steady-state simulations. In addition, the same system can be modeled by both terminating and nonterminating simulation models. The difference will lie in the parameters and the approaches taken in such studies. Nonterminating simulations in such cases will consider the long-run values of the relevant parameters. For example, if a bank is simulated by a terminating simulation model, the initial parameters of the simulation model may be empty queues and idle counters. This means that the bank is assumed to start a day with no customers at all. Now, this may happen in the case of a newly opened bank, a bank in a very remote place, or a bank that is not as reputed as other famous banks or bank branches. But for a very active and renowned bank, this scenario may not be applicable. In this case, the bank initially starts with the number of customers in a day and the initial number of customers is random. Thus, if the performance is measured by the delay in queue, then in the case of a terminating simulation, there may be no delay in the beginning. But for a nonterminating simulation, the delay may be experienced from the beginning. Thus, the estimated parameter for a nonterminating simulation will be the long-run expected delay in queue for a customer, whereas the performance measure for a terminating simulation may be the average delay in a queue of n number of customers based on the initial condition and parameter values.

Consider again a manufacturing system. It may be initialized by a prespecified number of parts in process. Thus, the system is initialized by fixed values, and the results of the simulation depend on this number of initial parts. The performance of this system may be measured by the expected delay in production. This is the performance measure for a terminating simulation. If a nonterminating simulation is applied, the system will have to run for a longer period to get rid of the effect of the initial parameter values. For a nonterminating simulation of such a system, the performance measure may be the expected long-run delay in production. However, there are numerous other real-life examples. What we really need to understand is that the type of simulation model to be applied depends on the type of system to be simulated. For example, if a system runs continuously in reality (e.g., for 24 hours), we need to apply a nonterminating simulation model. Otherwise, a terminating simulation model can be applied.

There are various approaches for the statistical treatment of the results of a simulation model. Statistics in terminating simulation may be based on the number of observations. Suppose D_i is the delay faced by the ith customer in a queue and d is the true value of the average delay in a queue. Then, for m number of customers, the average delay will be given by

$$\overline{D} = \frac{\sum_{i=1}^{m} D_i}{m}$$

In general, if Y_i is the measured performance of a parameter from a simulation run and its mean is given by $E(Y)$, then the confidence interval of n runs can be given by

$$\overline{X}(n) \pm t_{n-1,1-\alpha/2} \sqrt{\frac{s^2(n)}{n}}$$

In addition, there is another approach called the replication approach, in which independent simulation runs are replicated with various streams of random numbers although the initial conditions and the number of observations in each run are kept the same. Suppose that the number of observations at each run is denoted by n. There is another approach to apply a statistical analysis: measuring precision. They may be either absolute or relative measures. The hardcore mathematical representation of these measures is not discussed.

However, the advantage of a simulation study nowadays is that there is a vast number of simulation software applications available in the market. Every simulation software has built-in, easy-to-learn functions that perform statistical analyses of the results obtained from a simulation study. Thus, this basic understanding is expected to give the reader a good conception of the various aspects and types of simulation studies that can be conducted.

8.3 CONCLUSION

In this chapter, a glimpse of the analysis of the results of a simulation model has been given. Rather than going into the hardcore mathematical details of the concepts, a conceptual overview of the approaches to various simulation studies has been provided so that the reader can understand and identify the type of simulation study to be conducted based on the system under study.

REFERENCE

1. Law, A. M. and Kelton, W. D. (2003). *Simulation Modeling and Analysis*. 3rd Edition. New Delhi, India: Tata McGraw-Hill.

9 Computer Simulation

9.1 INTRODUCTION

Computer simulation is a computer process as well as a simulation tool. In Chapters 2 and 10 through 20, we introduce several aspects of simulation study, and in all cases, a number of software programs are shown in the respective fields of study. But before developing a computerized simulation model, there is always a stage of developing an idealized model. This idealized model helps in building the real simulation model.

In case of computer simulation, the visualization of the model by using a graphical user interface (GUI) is an important aspect. Moreover, visual simulation is a vast area of study. The elements of visual simulation are rendering, display modeling, model creation, and animation modeling. There are several basic languages that are used in this regard. In addition, there are general-purpose languages for simulations, such as MATLAB®, and languages under Visual Studio .NET. The other sets of languages, also known as object-oriented languages, include C++, Java, Python, Smalltalk, and so on. The main characteristics of object-oriented languages are inheritance, encapsulation, and polymorphism.

Simulation may be physical simulation, in which a small prototype of the actual larger system is built, or logic simulation, which generally uses a computer to implement the simulation model. Thus, logic simulations in all fields of study consist of built-in software systems used to simulate real-world phenomena. Simulation studies may be useful in concept building, providing open-ended experiences to the practitioners and tools for scientific methods, and helping in solving various real-world problems. Computer simulation may even be helpful in distance education. Computer simulation simulates several concepts in various fields of study, and accordingly, there are simulations in economics, chemistry, physics, biology, and so on.

Thus, computer simulation is also used for various purposes. However, the word "computer simulation" may indicate several aspects of simulation with a computer. Section 9.2 briefly discusses the various aspects or meanings of computer simulation systems.

9.2 COMPUTER SIMULATION FROM VARIOUS ASPECTS

As mentioned in Section 9.1, the word "computer simulation" may have several meanings or aspects. On the one hand, computer simulation may mean simulation of computer systems itself [1]. On the other hand, computer simulation may indicate the application of computerized simulation software in various fields of study. In addition, there are also other aspects, such as game simulation.

Simulation of a computer system means the simulation of various components of a computer. This includes simulation of functions of the central processing unit (CPU), simulation of memory functions, simulation of various electronic circuits in a computer system, and simulation of various programming system events in a computer.

As a second aspect, computer simulation may be used in various fields of study. This aspect particularly means the use of computerized simulation software in various disciplines. In almost all fields of study, simulation software is used widely. Examples of such fields include manufacturing, supply chains, mechanical engineering applications, civil engineering applications, electrical engineering applications, electronics, aerospace applications, and biological applications. Chapter 13 presents an overview of the application of various simulation software packages in various disciplines.

The third aspect is game simulation. Game simulation is a vast field of the application of simulation from an entirely different view. Thus, Section 9.5 is dedicated to game simulation. However, Section 9.3 provides an overview of simulation of computer systems.

9.3 SIMULATION OF COMPUTER SYSTEMS

A computer system is an assembly of several components, namely, CPU, memory, and input/output (I/O) system. At the lowest level of the design of a computer system, there are gate-level electronic circuits. Before discussing in depth the simulation possibilities of the various components of a computer, Section 9.3.1 presents a brief overview of them.

9.3.1 Various Components of Computer

The digital computer is an electronic machine. Its main capability is high-speed calculations. The basic structure of a computer consists of the following five fundamental blocks (Figure 9.1):

1. Arithmetic/logic unit (ALU)
2. Control unit (CU)
3. Memory
4. Input unit
5. Output unit

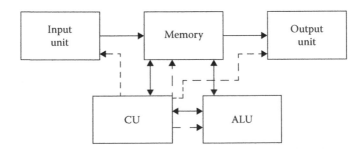

FIGURE 9.1   Basic structure of a computer. Dashed line, control signal; solid line, data signal.

The ALU contains the electronic circuits necessary to perform arithmetic and logic operations. The logic operations include COMPARE, ROTATE, AND, OR (will be explained later), and so on. The CU analyzes each instruction in the program and sends the relevant control signals to all the other units—ALU, memory, input unit, and output unit. A computer program, which contains instructions to make a computer to do a particular job, is fed into the computer through the input unit and is stored in the memory. To get the job done by the computer, the instructions in the program are fetched one by one from the memory by the CU. Then, depending on the task to be done, the CU issues control signals to the other units. After the task is accomplished (i.e., the instruction is executed), the result is stored in the memory or stored temporarily in the CU or ALU, so that it can be used by the next instruction of the program. After obtaining the final result of the problem, the result is shown by the output unit. The CU, ALU, and some registers (memory units with the CU) are collectively known as the CPU.

The memory is organized into locations. The smallest unit of location is called a bit. The number of bits in each location is known as a "word." Usually, the word size of memory is 8 bits, 16 bits, or 32 bits. In computer terminology, 8 bits together is called a byte. The total number of locations in memory is the memory capacity. Each memory location is identified by an address. The first location's address is 0. The number of bits required to specify the address depends on the capacity of the memory. If there are n address bits, it is possible to address 2 to the power of n (2^n) locations or words. For example, if there are 10 address bits, the maximum memory size is 1024 locations ($2^{10} = 1024$). In computer terminology, 1024 bits is taken as 1 kb. If each location is 8 bits wide, then the memory capacity is 1 kb.

9.3.1.1 Memory Types

Depending on the speed and volatility, the memory is of two types—primary or main memory and secondary or auxiliary memory. The primary memory is a large memory which is fast. It is also volatile in nature, that is, it loses its content when the power is off. This memory is directly accessed by the processor. It is mainly based on integrated circuits (ICs). The secondary memory is larger in size but slower than the main memory. It normally stores systems programs (programs that are used by the system to perform various operational functions), other instructions, programs, and data files.

Present-day memories use a semiconductor type of memory, which is of two types—static random access memory (SRAM) and dynamic random access memory (DRAM). SRAM preserves its content as long as the power supply is on, whereas each location in DRAM is refreshed every few milliseconds to retain its content.

Again, depending on the type of access to memory locations, the memory is of two types—RAM and sequential access memory (SAM). In RAM, the access times for all the locations are the same. It does not matter whether the location is topmost, lowermost, or intermediate. In SAM, as the name means, the access is sequential. Naturally, the time taken to access the first location is the shortest and the time taken to access the last location is the highest.

There are different types of storage devices apart from the main memory. These are called secondary storage devices, for example, floppy disks, magnetic disks, magnetic tapes, hard disks, and compact disks (CDs). Magnetic tape is a sequential memory. In floppy and hard disks, a combination of random and sequential access is followed.

Semiconductor memory is used in computers for two different applications:

1. Read/write memory (RWM)
2. Read-only memory (ROM)

In ROM, the memory can only be read and thus contains data permanently defined, so it cannot be written twice. There are different types of ROMs, some of which are rewritable. Some of the different types of ROMs are as follows:

ROM: It is programmed at the factory by the manufacturer just before shipment, so it becomes a permanent memory and its content cannot be erased.

Programmable ROM (PROM): It is a programmable device. The user buys a blank PROM and enters the desired contents using a PROM burner. It can be programmed only once and is not erasable.

Erasable programmable ROM (EPROM): It can be erased by exposing it to ultraviolet light for a duration of 6–14 minutes by an EPROM eraser.

Electrically EPROM (EEPROM): It is programmed and erased electrically. It can be erased and reprogrammed about 10,000 times. Any location can be erased and programmed.

Read/write memory: It is a type of memory that is easier to write into and read from compared to ROM.

The ALU contains hardware circuits to perform various arithmetic and logic operations. The hardware section of the ALU contains the following:

1. *Adder* adds two numbers and gives the result.
2. *Accumulator* is a register that temporarily holds the result of a previous operation in the ALU.
3. *General-purpose registers* (*GPRs*) store different types of information, such as operand, operand address, and constant.
4. *Scratch pad memory or registers* temporarily store the intermediate results of multiplication, division, and so on.
5. *Shifters and complementers* are also used for some arithmetic and logic operations.

The CU is the most complex unit of a computer. The main functions of the CU are as follows:

1. Fetching instructions
2. Analyzing and decoding instructions
3. Generating control signals for performing various operations

Some of the hardware resources of the CU are as follows:

1. Instruction register (IR)
2. Program counter or instruction address counter (IAC)
3. Program status word (PSW) register

There are several output devices, some of which are briefly mentioned in Section 9.3.1.2.

9.3.1.2   Monitors or Video Display Unit

Monitors provide a visual display of data. They look like televisions. They are of different types and have different display capabilities, which are determined by a special circuit called the adapter card. Some popular adapter cards are as follows:

- Color graphics adapter (CGA)
- Enhanced graphics adapter (EGA)
- Video graphics array (VGA)
- Super VGA (SVGA)

The smallest dot that can be displayed is called a pixel. The number of pixels that can be displayed vertically and horizontally gives the maximum resolution of the monitor. The resolution of the monitor determines the quality of the display. The higher the resolution, the better the quality of the display. Some popular resolutions are 800×640, 1024×768, and 1280×1024 pixels. Besides the video display unit (VDU), there are also printers and other output devices.

Secondary storage devices refer to floppy disks, magnetic disks, magnetic tapes, hard disks, CDs, and so on, which are used to store huge amounts of information for future use. The input unit, output unit, and secondary storage devices are together known as peripheral devices. A brief idea of some of the devices is given in Sections 9.3.1.3 through 9.3.1.5.

9.3.1.3   Floppy Disk Drives

In this device, the medium used to record the data is called a floppy disk. It is a flexible circular disk of diameter 3.5 inches made of plastic coated with a magnetic material. This is housed in a square plastic jacket. Each floppy disk can store approximately one million characters. Data recorded on a floppy disk are read and stored in a computer's memory by a device called a floppy disk drive (FDD). A floppy disk is inserted in a slot of the FDD. The disk is rotated normally at 300 revolutions per minute. A reading head is positioned touching a track. A voltage is induced in a coil wound on the head when a magnetized spot moves below the head. The polarity of the induced voltage changes when a 0 is read. The voltage sensed by the head coil is amplified, converted to an appropriate signal, and stored in the computer's memory.

- Floppy disks have the following capacities:
  - $5^{1/4}$ drive: 360 KB, 1.2 MB (1 KB = 2^{10} = 1024 bytes)
  - $3^{1/2}$ drive: 1.44 MB, 2.88 MB (1 MB = 2^{20} bytes)

9.3.1.4 Compact Disk Drives

A compact disk ROM (CD-ROM) uses a laser beam to record and read data along spiral tracks on a $5^{1/4}$ disk. A disk can store around 650 MB of information. CD-ROMs are normally used to store massive text data (such as encyclopedias), which are permanently recorded and read many times. Recently, CD writers have been available in the market. Using a CD writer, a lot of information can be written on a CD-ROM and stored for future reference.

9.3.1.5 Hard Disk Drives

Unlike a floppy disk, which is flexible and removable, the hard disk used in a personal computer (PC) is permanently fixed. The hard disk used in a higher end PC can have a maximum storage capacity of 17 GB (1 GB = 1024 MB = 2^{30} bytes). Nowadays, hard disk capacities of 540 MB, 1 GB, 2 GB, 4 GB, and 8 GB are quite common. The data transfer rate between the CPU and the hard disk is much higher compared to that between the CPU and the FDD. The CPU can use the hard disk to load programs and data as well as to store data. The hard disk is a very important I/O device. The hard disk drive (HDD) does not require any special care other than the requirement that one should operate the PC within a dust-free and cool room (preferably air-conditioned).

In summary, a computer system is organized with a balanced configuration of different types of memories. The main memory (RAM) is used to store programs that are currently executed by the computer. Disks are used to store large data files and program files. Tapes are serial access memories and used to back up the files from the disk. CD-ROMs are used to store user manuals, large text, and audio and video data.

9.3.1.6 Overall Method of Execution of Computer Programs

A computer program consists of both data and instructions for operations on data. The program is fed into the computer through the input unit and stored in the memory. To get the desired result, that is, to execute the program, the instructions are fetched from the memory one by one by the CU. Then, the CU decodes the instruction. Depending on the instruction, the CU sends signals to the I/O unit or memory, and accordingly the corresponding unit does whatever is instructed by the CU. In this way, instructions get executed. After an instruction is executed, the result of the instruction is stored in the memory or stated temporarily in the CU or ALU so that it can be used by the next instruction. The results of a program are taken out of the computer through the output unit.

9.3.1.7 Software Used in Computers

A computer contains two parts—hardware and software. Hardware is the physical components that make up a computer, and software is the set of instructions that perform a particular task to provide the desired output. A computer software consists of a set of instructions that mold the raw arithmetic and logic capabilities of the hardware units to perform a task.

The process of software development is called programming. To do programming, one should have a knowledge of (1) a particular programming language and (2) a set of procedures (algorithms) to solve a problem or develop a software. The development of an algorithm is basic to computer programming. Developing a computer program is a detailed process that requires serious thought, careful planning, and accuracy.

Computer software can be broadly classified into two categories—system software and application software. System software is a set of system programs that are required to run other programs and user applications. Thus, system programs are designed to make the computer easier to use. An example of system software is an operating system, which consists of many other programs for controlling the I/O devices, memory, processor, and so on. To write an operating system, the programmer needs instructions to control the computer's circuitry (hardware part), for example, instructions that move data from one location of storage to a register (a unit of storing data) of the processor.

Application software is a set of application programs that are designed for specific computer applications, such as payroll processing and library information management. To write programs for payroll processing or other applications, the programmer does not need to control the basic circuitry of a computer. Instead, the programmer needs instructions that make it easy to input data, produce output, do calculations, and store and retrieve data. There are many languages available for developing software programs. These languages are designed keeping in mind some specific areas of applications. So, some programming languages may be good for writing system program or software and some others for writing application software.

9.3.1.8 Computer Language

A computer executes programs only after they are represented internally in binary form (sequences of 1s and 0s). Programs written in other languages must be translated into binary representations of the instructions before they can be executed by the computer. Programs written for a computer may be in one of the following languages:

1. *Machine language or first-generation language (1GL).* This is a sequence of instructions written in the form of binary numbers consisting of 1s and 0s to which the computer responds directly. Just as hardware is classified into generations, computer languages also have generation numbers depending on the level of interaction with the machine. Machine language is considered to be the 1GL. It is faster in execution since the computer directly starts executing it, but it is difficult to understand and develop.

2. *Assembly language or second-generation language (2GL).* When symbols (letters, digits, or special characters) for the operation part, the address part, and other parts of the instruction code are employed, then the representation is called an assembly language. Both machine and assembly languages are referred to as low-level languages since the coding for a problem is at the individual instruction level. An assembler is required to translate an assembly language program into an object code executable by the machine, which is considered to be the 2GL. Writing programs in assembly language is easier than in machine language. But this language is specific to a particular machine architecture; that is, each machine has its own assembly language that is dependent upon the internal architecture of the processor.

3. *High-level language or third-generation language (3GL).* In this language, one can specify in detail not only what one wants but also how it is to be

achieved. For this reason, these languages are called procedural languages. A compiler translates a high-level language into an object code (machine-understandable form). The time and cost of creating a high-level language program are much less than that of the 1GL and 2GL. High-level language programs are translated into the machine-understandable form (object code) by a software called a compiler. There is another type of translator software called an "interpreter." Compilers and interpreters have different approaches to translation. High-level languages are more readable than the 1GL and 2GL, they may be run on different machines with little or no change, errors in the programs can be easily removed, and they are easy to develop, for example, C, FORTRAN, and BASIC.

4. *Fourth-generation language (4GL)*. It is a nonprocedural language. Major 4GLs are used to get information from files and databases (will be explained later). These languages are machine independent. They are designed to be easily learned and used by end users. Examples include ORACLE and structured query language (SQL).

5. *Fifth-generation language (5GL)*. It emphasizes the concept of artificial intelligence.

9.3.1.9 System Software

These are general programs written for the system, which provide the environment to facilitate the writing of application software. Some system programs are given as follows:

1. *Language translator*. Computers cannot understand the English language or characters or symbols. They can only understand signals. A digital signal consists of 1s and 0s as the two signal levels. A language translator translates a computer program (consisting of instructions to make the computer do a particular job) written by the user into a machine-understandable form (consisting of 1s and 0s). Some language translators are given as follows:

   Compiler—It is a translator system program used to translate a high-level language program into a machine language program.

   Assembler—It is a translator system program used to translate an assembly language program into a machine language program.

   Interpreter—It is a translator system program used to translate a high-level language program into a machine language program, but it translates and executes line by line.

2. *Operating system*. It is the most important system software required to operate a computer system. It manages the system resources, takes care of the scheduling jobs for execution, and manages the flow of data and instructions between the I/O units and the main memory, for example, DOS, Windows (different versions), and UNIX. There are different types of operating systems—batch operating systems, multiprogramming operating systems, network operating systems, and distributed operating systems. The main functions of operating systems as a whole are information management, process management, and memory management.

3. *Utilities.* They are those that are very often requested by many application programs. Examples are SORT programs for sorting large volumes of data.

4. *Special purpose software.* They are the programs that extend the capability of operating systems to provide specialized services to application programs, for example, database management software, such as SQL, Microsoft Word, and Microsoft Excel.

5. *Loader.* It is a system program used to store the machine language program into the memory of the computer.

6. *Linker.* During the linking process, all the relevant program files and functions are put together that are required by the program. For example, if the programmer is using the pow() function in the program, then the object code of this function should be brought from a library file of the system and linked to the main program.

9.3.1.10 Elements of Programming Language

There are several elements of a programming language. All the elements together describe the constructs of a programming language. The basic elements are as follows:

Variable. It is a character or group of characters assigned by the programmer to a single memory location and used in the program as the name of that memory location to access the value stored in it. For example, in the expression A = 5, A is the name of a memory location, that is, where 5 is stored.

Constant. It has a fixed value in the sense that 2 cannot be equal to 4. There can be integer, decimal, or character constants.

Data type. There are different types of data processed by the computer—numbers and words. Data types define sets of related values or integers, numbers with fractions (decimals), characters, and sets of specific operations that can be performed on those values.

Array. It is a collection of the same type of data (either string or numeric), all of which are referenced by the same name.

Expression. It represents an algebraic expression depicting the arithmetic and logic operations among the variables considered. We know that we can express intended arithmetic operations using expressions such as $x + y$ and $xy + z$.

I/O statements. These are instructions that are used to instruct the computer to perform the input operations that are required for a program to provide the desired output and the output operations that are required to instruct the computer to produce and show the output as directed.

Conditional and looping statements. If the execution of an instruction depends on some condition, then such an instruction is called a conditional statement. Similarly, if a set of instructions is to be performed iteratively, then such instructions are called looping statements.

Subroutines and functions. If there is/are one or more tasks that are to be performed to get the desired result at different places of a program or if there is/are one or more tasks that are required to be performed separately for some purpose, then the particular set of instructions for that task is grouped into a function or subroutine. Although functions and subroutines are used for the mentioned purposes, there are differences between these two.

9.3.2 SIMULATION OF VARIOUS COMPONENTS OF COMPUTER SYSTEMS

The simulation of a computer system means the simulation of all the components discussed in Section 9.3.1. In addition, the entire computer system is run by electricity, and thus electrical properties of the circuits can also be simulated. The main purpose of simulating the electrical properties include checking of signal timings, various electrical parameter values related to these circuits, and logic behavior of the circuits. There are various low-level languages that are used to measure some of the functioning of the various components, such as assembly language programming. However, there are other languages for various purposes.

Various software tools are available to simulate computer systems. One of the main characteristics of simulation tools is model building support. Toward this end, there are various software programs that can do the job. Such tools include very-high-speed IC (VHSIC) hardware description language (VHDL), SimPack, and CSIM. Among all these tools, the basic characteristics of VHDL that came into action in 1980 are given as follows:

1. It is used to simulate microelectronic circuits.
2. It is a standard for electronic system simulators defined by the Institute of Electrical and Electronics Engineers (IEEE) and the US Department of Defense.
3. It facilitates hardware modeling from gate level to system level.
4. Its features facilitate digital design and reusable design documentation.
5. It can be used for the performance prediction of electronic circuits.
6. The models developed in VHDL may be divided into three domains— functional domain, structural domain, and geometric domain.
7. It allows both top-down and bottom-up approaches of design.
8. The features of the VHDL tool allow users to simulate various features of a circuit, such as behaviors of a circuit over time, functions of a circuit, signal timings of a circuit, structure of a circuit, and concurrency of circuit operations.
9. It helps to explore design alternatives.
10. It helps in reuse of designs among several vendors' tools.
11. The components of a VHDL model include entity declarations, architecture descriptions, and a timing model.
12. An entity in a VHDL model represents the I/O specification of the components.
13. The architecture in a VHDL model defines the internal structure of the entities in the model.
14. A design entity in a VHDL model may represent the design of an IC, a printed circuit board (PCB), or a gate.
15. It uses data types, objects, attributes, sequential and concurrent statements, function libraries, operators, packages, ports, generics, buses, and drivers to define a model.

16. It is independent of any other technology.
17. Various types of architectures may be represented by a VHDL model. The first type of architecture may define the subcomponents and the connections among them. The second type of architecture may define the concurrent statements. The third type of architecture may define the behavioral descriptions.

In addition, other tools and simulators are available. For example, CPUSimul software can simulate the following four components of a computer:

1. Data are fetched from memory and written onto memory by a drag-and-drop option provided by the software. The operations that are required to be performed can be selected by the user.
2. There is a provision for an accumulator machine program. Assembly language is used to facilitate the programming. Assembly language source files have the extension .amp.
3. There is also a provision for simulating a stack-based machine. The relevant assembly language programming files have the extension .smp.
4. General-purpose registers can also be simulated by using assembly language programming. The related assembly language programming files have the extension .gmp.

The above-mentioned machine types can be simulated by CPUSimul. The user chooses any of these machine types and the configuration file stores the size of the small memory that is required for the selected machine. The size of the memory can be increased depending on the requirement of the user.

The registers that are simulated are those that cannot be programmed by the user. These registers are required for other programs. Some of these registers that are simulated are as follows:

- IR points to the instruction that will be executed next.
- Memory address register contains the address of the memory cell either from which the data are fetched or to which the data are written.
- Memory buffer register contains data that are to be written to the memory or fetched from the memory.
- Program counter points to the current instruction in execution.
- Status register contains four flags.
  - Flag zero is set to 1 if the result of a logic or arithmetic operation is 0.
  - Flag negative is set to 1 if the value of the most significant bit is 1.
  - Flag carry is set to 1 if an overflow occurs for an arithmetic operation.
  - Flag overview is set to 1 in case of an algebraic overflow.

Memory can also be simulated in various ways. An example is the simplified virtual memory (VM) simulation that shows the basic concepts of VM paging. The applet starts off with eight processes that take turns referencing memory locations. A certain color indicates that the process is the running one, and another color indicates that the process is ready to run and is waiting for its turn. There are 16 VM pages and 8 translation lookaside buffer (TLB) page table entries. Initially, the TLB page table entries are white, indicating that the page "slot" is empty. As the simulation progresses, the running process randomly chooses a page. If the page is a TLB miss, the process changes color to a third color (indicating that a page fault has occurred), and the requested page is brought into the TLB replacing an empty slot (if one is available) or the least recently used (LRU) page in the TLB. All "new" pages in the TLB are colored in a fourth color. If the page is a TLB hit, then the process can either read the page contents or write to it. If the process writes to the page, then the page changes color to a fifth color, indicating that it is dirty and should be copied back to disk before being replaced by another page. At all times, there are lines connecting the TLB pages to the VM pages they are mapped to and also a line connecting the running process to the TLB page it is accessing.

The memory system of a computer can also be modeled by a VHDL tool. In its simplest form, memory can be thought of as a contiguous array of bits. Thus, simulating such an array is interesting from a programming point of view. Thus, memory can be simulated by the use of arrays. In addition, various data structures can also be used to simulate the operations performed in the memory. For example, the data structure called stack can be used for those parts that use last-in, first-out (LIFO) policy. In addition, various memory operations, print instructions from computers, and I/O operations also use a different data structure known as "queue." The basic concepts of stacks and queues are given in Sections 9.3.2.1 and 9.3.2.2.

9.3.2.1   Stack

A basic structure of a stack is shown in Figure 9.2. In a stack, the items can be fetched from or inserted on the top only. For this reason, a stack is also called a LIFO data structure. There are only two operations performed on a stack:

- Push: To insert data on the top of a stack
- Pop: To fetch data from the top of a stack

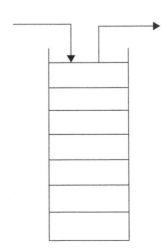

FIGURE 9.2   Diagrammatic representation of a stack.

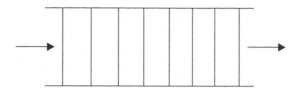

FIGURE 9.3 Diagrammatic representation of a queue.

Thus, to fetch data that are not located on the top, the other data will have to be fetched until the particular data to be fetched are on the top. Then, the particular data are fetched and the previously fetched data are again pushed on to the top of the stack in the reverse order. Stack is used while calling functions and using recursion. Recursion is an algorithm in which a function calls itself.

9.3.2.2 Queue

The structure of a queue is shown in Figure 9.3. From the figure, it is clear that a queue is a first-in, first-out (FIFO) structure in which items that enter first will exit first. Such structures are used to simulate the queues of printers, memory units, various processes, and so on.

One of the two ends of a queue is called the "front" and the other end of the queue is called the "rear." Items are appended to the rear of the queue and the items are fetched from the front of the queue.

9.4 COMPUTER SIMULATIONS FOR VARIOUS FIELDS OF STUDY

The application of computerized software for simulation purposes is another aspect of computer simulation. Chapter 13 describes in more detail the application of simulation in various fields of study. However, it can be briefly discussed that there are numerous disciplines in which computerized simulation software are used. Both open-source software and proprietary software are used in the fields of electrical engineering, electronics engineering, mechanical engineering, civil engineering, computer engineering, aerospace engineering, bioengineering, biotechnology engineering, oceanology, and so on. In addition, there are behavioral simulation software such as system dynamics software, whose applications lie in various disciplines and various sectors of management.

Computerized simulation software programs can also be used in various network studies. These software programs are commonly known as network simulators. Network simulation particularly means simulation of network algorithms. It always needs open platforms since various types of packages are used in a network. Thus, network simulation should have the flexibility to allow different types of packages to run transparently without affecting the existing packages in the network.

Network simulators are used to simulate various network protocols. In general, a network simulator comprises various technologies and protocols that facilitate the simulation of complex networks. There are a plenty of both commercial software and open-source software in this area. Examples of commercial software include OPNET and QualNet, whereas examples of open-source software include network simulator, J-Sim, OMNeT++, and SSFNet. Among these software programs, OPNET is a very flexible software that supports object modeling, grid computing, wireless modeling, kernel simulation, discrete event simulation, analytic simulation, hybrid simulation, and GUI-based debugging and analysis.

9.5 GAME SIMULATION

Game simulation or simulation game or simply game is a simulation of various activities in daily life for the purpose of playing games. There are no strict rules for these simulations. Popular types of such games include war games, role-playing games, and business games. These games are very efficient and effective applications of artificial intelligence, and they all make a very impressive use of animation for visual simulation of daily activities.

Game simulations have a plenty of examples. The most remarkable ones may include flash games, action games, strategic games, sports games, and adventure games. The learning rates for these games vary among various ages, genders, and cultures.

Complex visual programming skills are always required for developing these games. The real-life events are developed; every game is analyzed thoroughly before it is developed and an in-depth animation skill is required for such developments. Some of the popular software categories for game simulations are listed below based on the various types of modeling for these kinds of simulations:

1. 3D modeling software, e.g., OPAL, Cafu, CityEngine, Delta3D, Open-Simulator, Open CASCADE, and Visual3D Game Engine
2. Aerodynamics modeling, e.g., APAME and E-AIRS
3. Agent-based simulation, e.g., Lapsang and MASON
4. Discrete event simulation, e.g., JaamSim, Tortuga, and GeneSim

There are also several other categories. However, some of the popular types of game simulations based on different real-life events are given as follows:

- Biological simulation
- Social simulation
- Business simulation
- City building
- Government simulation
- Sports game simulation, such as baseball, basketball, cricket, football, hockey, racing, and rugby
- Vehicle simulation including bus, truck, train, flight, ship, racing, and submarine
- Trade simulation
- Photo simulation
- Medical simulation

9.6 CONCLUSION

In this chapter, an overview of the various computer simulations has been presented. Computer simulation may indicate various computerized software applications for various fields of study or simulation of computer systems. Various other aspects can also be observed in this area. One such significant aspect includes game simulation or simply game. Computer games for simple entertainment can be thought of as a kind of simulation, since these software applications simulate daily activities through animation and complex programming using various software applications. Thus, this chapter has also provided an overview of game simulation.

However, computer simulation has a vast scope for improvement. The sophistication of computerized simulation is a result of the advent of various technologies. Virtualization plays a big role in computer simulation. It can be implemented by the latest technologies such as cloud computing.

REFERENCE

1. Banks, J., Carson II, J. S., Nelson, B. L., and Nicol, D. M. (2003). *Discrete-Event System Simulation*. 3rd Edition. New Delhi, India: Prentice Hall.

10 Manufacturing Simulation

10.1 INTRODUCTION

Manufacturing systems are methods of organizing production. Many types of manufacturing systems are in place, including assembly lines, batch production, computer-integrated manufacturing, and flexible manufacturing systems (FMSs).

Manufacturing stems from the Latin words *manus* (hand) and *factus* (make). In sum, manufacturing means the process of converting a raw material into a physical product. A system is a collection of elements forming a unified whole. Thus, a manufacturing system refers to any collection of elements that converts a raw material into a product. We can identify and categorize manufacturing systems as given below:

- At the *factory level*, the manufacturing system is the factory (also referred to as a plant, a fabricator, or simply fab). The elements of the system are areas and (groups of) machines.
- At the *area level*, the manufacturing system is an area of the factory with several machines or groups of machines (cells). The elements of the system are individual machines.
- At the *cell level*, the manufacturing system is a group of machines that are typically scheduled as one entity.
- At the *machine level*, the manufacturing system is the individual machine (also referred to as equipment or tools). The elements of the system are machine components.

To analyze and optimize a manufacturing system, we need a way to model it and to determine its performance. We can make many different types of models. Often it is possible to make a rough estimation of the plant performance without requiring advanced modeling. In the early stages of the design process, we often could not do better than rough estimates, since we do not have enough detailed data. As we collect more data, it is possible, for simple plants, to do some more accurate calculations with simple planning. However, this planning has a limited range of applications and validity. If plants are more complex, we require better planning that incorporates uncertainties. However, these models require extensive mathematical knowledge and substantial effort, and they have a limited range of applicability.

There are numerous aspects of manufacturing systems. The current research study has focused its application on mainly parallel machine scheduling and routing of jobs. Thus, a discussion on scheduling is presented in Section 10.2.

10.2 SCHEDULING

A schedule is a tangible plan or document such as a bus schedule or a class schedule. Scheduling is the process of generating a schedule. In the first step, we plan a sequence or decide how to select the next task. In the second step, we plan the start time and the completion time of each task. Three types of decision-making goals can be prevalent in scheduling—turnaround, timeliness, and throughput. Turnaround measures the time required to complete a task. Timeliness measures the conformance of a particular task's completion to a given deadline. Throughput measures the amount of work completed during a fixed period of time.

If a set of jobs available for scheduling does not change over time, the system is called a static system; in contrast, if new jobs appear over time, the system is called a dynamic system. When the conditions are assumed to be known with certainty, the model is called a deterministic model. In contrast, when we recognize uncertainty with explicit probability distributions, the model is called a stochastic model. Two kinds of feasibility constraints are commonly found in scheduling problems. First, there are limits on the capacity of the machines, and second, there are technological restrictions on the order in which some jobs can be performed. A scheduling problem gives rise to allocation decisions and sequencing decisions.

The single machine problem is fundamental in the study of sequencing and scheduling. It is considered as a simple scheduling problem because it does not have distinct sequencing and resource allocation dimensions. Challenging optimization problems are encountered even in the simplest of scheduling problems. Some of the traditional methodologies to solve such problems include dynamic programming and branch-and-bound methods.

Parallel machine scheduling is a very traditional branch of research studies. The scheduling concept here consists of the job sequencing problem for arranging jobs in a sequence over time. The research studies in the existing literature have investigated the parallel machine scheduling problem from various aspects with an endeavor to minimize makespan, tardiness, completion time, waiting time, idle time, and so on. In job sequencing, finding the right sequence to satisfy the above constraints for multiple jobs and multiple machines is still a challenging research study. Numerous research efforts are observed toward this direction.

A vast number of job sequencing problems have been investigated by the researchers, and sometimes, similar sets of objectives are observed in more than one research study. The various methods to solve this type of problem include both deterministic and nondeterministic methods, such as linear programming, dynamic programming, integer programming, genetic algorithm, particle swarm optimization, ant colony optimization, and simulated annealing.

Manufacturing systems have always been the prime focus of simulation studies. Although the word "simulation" is widely used nowadays, even in computer applications, animations, and other fields of study, the manufacturing field is still the prime concern of simulation. Most of the books on simulation have emphasized manufacturing simulation. In addition, numerous research publications are observed in this area, for example, in the ScienceDirect, Springer, Taylor & Francis, IEEE, Emerald, JSTOR, and other famous international databases. Thus, manufacturing simulation is a very traditional branch of simulation. As a result, numerous aspects of manufacturing systems have simulation as evident from the existing literature. This chapter is an endeavor to give a glimpse of this vast area of research in simulation.

The remaining sections of this chapter are arranged in the following order: Section 10.3 shows various aspects of manufacturing systems that have been simulated in the existing literature. Section 10.3.8 depicts the general methodology of selecting various simulation software applications for manufacturing. Section 10.3.9 lists some of the most significantly used simulation software for manufacturing. Section 10.4 provides the conclusion for this chapter.

10.3 ASPECTS OF MANUFACTURING FOR SIMULATION STUDY

The existing literature shows the various aspects of manufacturing systems that have been simulated till today. These aspects of research are recorded [1] and the main areas on manufacturing simulation have been confined into the following:

1. Manufacturing systems design
2. Manufacturing systems operation
3. Development of manufacturing simulation software

Various prime subareas that have been considered for simulating manufacturing systems designs are as follows:

1. Design of facility layout
2. Design of material handling systems
3. Design of cellular manufacturing systems
4. Design of FMSs

Various prime subareas that have been considered for simulating manufacturing systems operations are as follows:

1. Operations scheduling
2. Operating policies
3. Performance analysis

Various prime subareas that have been considered for developing simulation software for manufacturing are as follows:

1. Selection of simulation software for manufacturing
2. Proposing new simulation software

Each of these aspects has several characteristics that are considered while simulating those systems. The characteristics for each of these aspects are presented in Sections 10.3.1 through 10.3.7.

10.3.1 ASPECTS CONSIDERED FOR DESIGN OF FACILITY LAYOUT

The characteristics or aspects considered for the design of a facility layout may include the following:

1. Facility planning
2. Process layout
3. Product layout
4. Fixed position layout
5. Mixed or hybrid layout
6. Cellular layout
7. Intermittent processing system
8. Continuous processing system
9. Various resources used
10. Scheduling issues
11. Work flow
12. Group technology
13. Layouts of various types of facility, such as warehouse, shop floor, and layout
14. Various variables influencing layout
15. Space, total budget, and location-related factors
16. Line balancing

10.3.2 ASPECTS CONSIDERED FOR DESIGN OF MATERIAL HANDLING SYSTEMS

The aspects considered for the design of material handling systems include the following:

1. Types of material handling systems, such as conveyors, automated guided vehicles, robots, cantilevered cranes, elevators, pallets, and trucks
2. Various types of storage systems, such as pallet storage, racks, and buffers
3. Automated storage and retrieval systems
4. Scheduling of material handling equipments
5. Routing of material handling equipments
6. Dispatch policies
7. Loading and unloading of machines
8. Buffer management
9. Unconstrained vehicles

10.3.3 Aspects Considered for Design of Cellular Manufacturing Systems

The aspects considered for the design of cellular manufacturing system include the following:

1. Cell formation
2. Uncertainty issues in cellular manufacturing
3. Managing clusters in cellular manufacturing
4. Dynamic system configuration
5. Layout design
6. Lot sizing
7. Route selection
8. Part-machine assignment problem
9. Material handling selection among cells
10. Formation of part families
11. Intercell communication
12. Grouping of machines

10.3.4 Aspects Considered for Design of FMSs

The aspects considered for the design of FMSs include the following:

1. Machine flexibility
2. Various changes in the system
3. Dynamic system configuration
4. Automation of various aspects
5. Inventory control
6. FMS planning
7. Machine scheduling
8. FMS communication protocol
9. Material handling system in FMS

10.3.5 Aspects Considered for Operations Scheduling

The aspects considered for operations scheduling include the following:

1. Flow shop scheduling
2. Job shop scheduling
3. Assembly operations
4. Quay crane scheduling
5. Single machine scheduling
6. Parallel machine scheduling
7. Various timing considerations, such as completion time, makespan, and setup time

8. Scheduling of batch operations
9. Dynamic scheduling
10. Scheduling of cross-dock operations
11. Operator scheduling
12. Scheduling in various applications or various fields of applications and so on

10.3.6 ASPECTS CONSIDERED FOR OPERATING POLICIES

The aspects considered for operating policies for simulation include the following:

1. Scheduling shifts
2. Maintenance
3. Packing
4. Shipping
5. Routing of parts
6. Processing characteristics
7. Rework of parts
8. Quality inspection
9. Customer order processing
10. Storage systems
11. Buffering systems
12. Tooling and fixtures
13. Capacity planning
14. Inventory control
15. Purchasing system
16. Shop floor control
17. Vendor management
18. Loading and unloading operation at various places and so on

10.3.7 ASPECTS CONSIDERED FOR PERFORMANCE ANALYSIS

The aspects considered for performance analysis are as follows:

1. Total cost
2. Total inventory
3. Service level
4. Delivery schedule
5. Customer feedback
6. Customer satisfaction
7. Return on investment
8. Delivery delays
9. Manufacturing lead time
10. Order lead time
11. Conformance to specifications
12. Transportation cost
13. Transport delay and numerous others

Thus, all these aspects should be considered while developing simulation models in manufacturing systems, as evident from the existing literature.

10.4   SELECTION OF SIMULATION SOFTWARE

A significant number of research papers are observed in the existing literature depicting how to select simulation software. There are various approaches and a number of factors to consider when selecting simulation software for simulating manufacturing systems [2].

Among the various factors mentioned in the existing literature, the most significant ones include the following:

1. Consideration of available funds and other resources
2. Willingness of management to purchase simulation software
3. Future applications of the simulation software to be purchased
4. Risk involved
5. Lead time considerations
6. Training cost
7. Learning capability of existing employees
8. Reduction of capital costs
9. Complexity involved in the operations to be simulated
10. Availability of associated software
11. Resources required
12. Total budget
13. Discrete or continuous simulation required
14. Time required to implement the system
15. Education level of the existing employees

However, the factors to consider when selecting simulation software vary among organizations. Based on these factors, an overall methodology can be decided which can be used to choose a simulation software for simulating manufacturing systems. A number of research studies are observed in the existing literature proposing methods for selecting simulation software for manufacturing systems. However, Hlupic and Paul [3] have proposed a generalized approach to select simulation software, which is shown in Figure 10.1. It is a generalized method to make a decision to purchase any software including simulation software. It starts with investigating the existing resources that are currently present in an organization, followed by a survey of the available software in the market. A gap analysis is required to identify the gap between what is available on hand and what is required if a software is purchased. The cost and utilities of all the existing software applications should be performed in light of the available budget and the resources available. After evaluating all the software applications, the purchasing decision can be made.

Section 10.3.9 shows a list of the most used software applications in the field of manufacturing simulation.

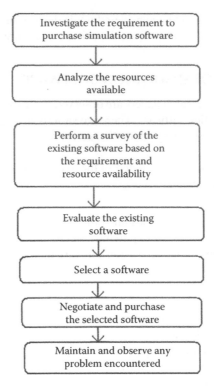

FIGURE 10.1   A generalized method of selecting simulation software.

10.5   LIST OF SIMULATION SOFTWARE APPLICATIONS

This section provides a list of frequently used software for simulating manufacturing systems. Some of these software applications are also flexible enough to be used in both the manufacturing and service industries. Most of these software applications use sophisticated visual tools so as to facilitate the visual presentations of simulation studies. A list of such software applications along with their vendors is provided in Table 10.1.

10.5.1   INTRODUCTION TO ARENA SIMULATION SOFTWARE

This section discusses modeling with simulation. Arena (version: 7) has been used for building models. Each of the components will be presented with a detailed description. Before going into further detail, we must first introduce each of the design components as given in the following paragraphs. The operations for each of the components could not be shown because of the limitation of the software (student version) used.

The fundamental building blocks of Arena are called *modules*. There are two types of modules in Arena: flowchart modules and data modules.

The *flowchart modules* describe the dynamic processes in the model. The *data modules* define the characteristics of the various process elements, such as entities and resources.

TABLE 10.1

List of Most Significant Software Applications for Manufacturing Simulation

Software	Vendor	Typical Applications
@Risk	Palisade Corporation	Risk analysis and decision making under uncertainty
Analytica	Lumina Decision Systems Inc.	Investment, risk, decision, portfolio, and network flow analysis; systems dynamics; resource and R&D planning; and organization simulation
AnyLogic	XJ Technologies	Marketplace and competition modeling, supply chains, logistics, business processes, project and asset management, pedestrian dynamics, and health economics
Arena	Rockwell Automation	Facility design/configuration, scheduling, effective passenger and baggage-handling processes, patient management, and routing/dispatching strategy
AutoMod	Brooks Software	Decision support tool for the statistical and graphical analysis of material handling, manufacturing, and logistical applications using true to scale 3D graphics. Templates for conveyor, path-based movers, bridge cranes, automated storage and retrieval system (AS/RS), Power & Free, and kinematics
Crystal Ball	Decisioneering	Planning, cost, and benefit analysis; risk and project management; petroleum exploration; and Six Sigma
eM-Plant	UGS	Object-oriented, hierarchical discrete event simulation tool for modeling visualization, and planning and optimization
Enterprise Dynamics Simulation Software	Modeling Corporation	Material handling, manufacturing, call center, and service industry applications; process improvement; and capacity planning
Extend OR	Imagine That Inc.	Message-based discrete event architecture, models, and processes involving physical or logical objects
FlexSim	FlexSim Software Products Inc.	Manufacturing, logistics, material handling, container shipping, warehousing, distribution, mining, and supply chain
Micro Saint Sharp Version 2.1	Micro Analysis & Design	General-purpose, discrete event simulation modeling environment. Improves facilities design, maximizes worker performance, and so on
ProModel Optimization Suite	ProModel Corporation	Lean Six Sigma, capacity planning, cost analysis, process modeling, cycle time reduction, throughput optimization, and so on
SIMPROCESS	CACI Products Company	Process simulation, process improvement, optimization, business process management (BPM), business activity monitoring (BAM), dashboards, capacity planning, logistics and supply chain, inventory operations, airport operations, hospital operations, insurance, financial operations, and manufacturing processes

The panels in the Arena window consist of the building blocks that are used to design models. There are different types of panels, which are as follows:

1. *Basic process panel*. It is used for fundamental model building and consists of eight flowchart modules and six data modules.
2. *Advanced process panel*. It consists of smaller building blocks for more detailed modeling.
3. *Advanced transfer panel*. It consists of modules for moving (transferring) entities around.
4. *Blocks and elements panels*. It contains the modules that give full access to the SIMAN simulation language that underlies Arena.

There are several different kinds of modules. Some of those that will be used in our model are introduced in Sections 10.5.1.1 and 10.5.1.2.

10.5.1.1 Flowchart Modules

1. *Create flowchart module*. This module is used for creating a new entity to the system shown in Figure 10.2. The detailed window (Figure 10.3) of this module can be obtained by double-clicking on the module shown in Figure 10.2. Here, we enter the name of the Create module in the Name textbox, the type of the entity created or arrived, the time between arrivals,

FIGURE 10.2   Create module.

Create			? X
Name:		Entity Type:	
Create 1	▼	Entity 1	▼
Time Between Arrivals			
Type:	Value:	Units:	
Random (Expo) ▼	1	Hours	▼
Entities per Arrival:	Max Arrivals:	First Creation:	
1	Infinite	0.0	
	OK	Cancel	Help

FIGURE 10.3   Create module window.

the entities per arrival, the maximum number of entities arrived, and the time of the first arriving entity.

2. *Process flowchart module.* This module is the main processing method in the simulation model. The respective design symbol is shown in Figure 10.4. The detailed window of this module is shown in Figure 10.5.

    Here, we enter the name of the process (Name), the type of the process (Type), and the action taken by the process—Delay, Seize Delay, Seize Delay Release, and Delay Release. For waiting only, we can use the action "Delay," and for holding (seizing), processing (Delay), and the releasing resource after use (Release), we use the action "Seize Delay Release." We also add the resource and enter the processing time (Delay Type).

3. *Assign flowchart module.* This module is used for assigning new values to variables, entity attributes, entity types, entity pictures, or other system variables shown in Figure 10.6. The detailed window of this module is shown in Figure 10.7.

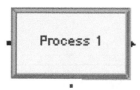

FIGURE 10.4   Process module.

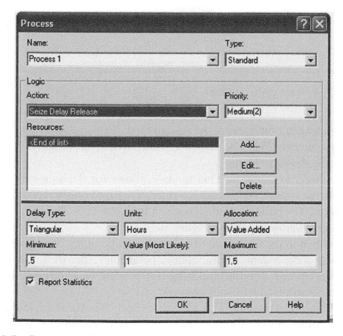

FIGURE 10.5   Process module window.

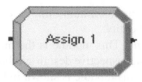

FIGURE 10.6   Assign module.

FIGURE 10.7   Assign module window.

Here, we enter the name of the Assign module (Name), the variables, and their assignments. When we click the "Add" button, an Assignment window appears as shown in Figure 10.7.

4. *Decide flowchart module.* This module allows for decision-making processes in the system. We can branch from here based on some condition or a percentage. The building module is given in Figure 10.8. The corresponding data input window is given in Figure 10.9.

   As shown in the figure, we can enter the name in the Name field. If the branching is dependent on some chance percentage, then the respective percentage is entered, or if the branching is dependent on some condition, then the corresponding condition is entered.

5. *Record flowchart module.* This module collects statistical data for analyzing the Arena model. The building module and the detailed window are shown in Figures 10.10 and 10.11, respectively.

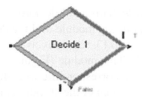

FIGURE 10.8   Decide module.

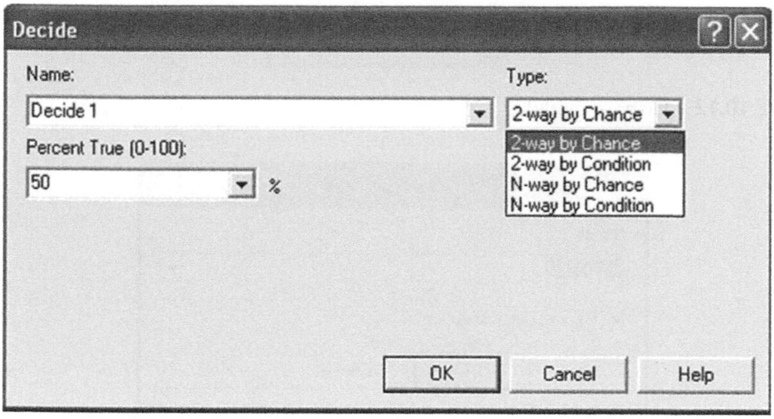

FIGURE 10.9   Decide module window.

FIGURE 10.10   Record module.

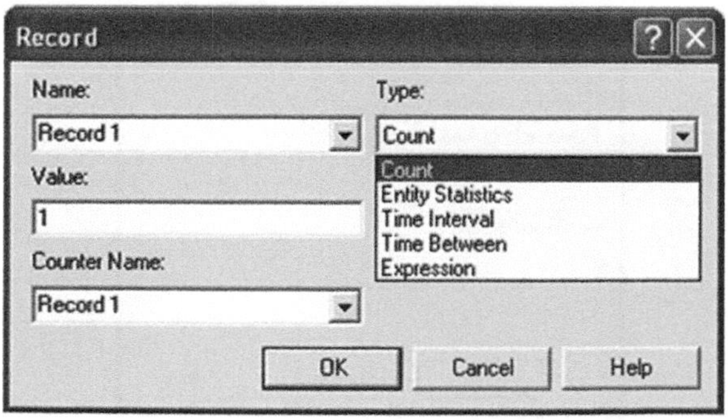

FIGURE 10.11   Record module window.

6. *Dispose flowchart module.* This module indicates the ending point in a simulation model. The building module and the detailed input window are given in Figures 10.12 and 10.13, respectively.

7. *Station flowchart module.* This module (Figures 10.14 and 10.15) defines a station (or a set of stations) corresponding to a physical or logical location where processing occurs. We enter the station name in the Name field.

FIGURE 10.12   Dispose module.

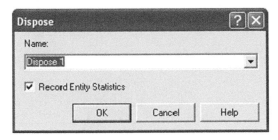

FIGURE 10.13   Dispose module window.

FIGURE 10.14   Station module.

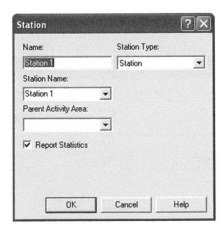

FIGURE 10.15   Station module window.

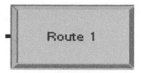

FIGURE 10.16   Route module.

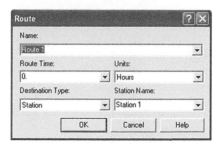

FIGURE 10.17   Route module window.

8. *Route flowchart module.* This module (Figures 10.16 and 10.17) transfers an entity to a specific station or to the next station in a sequence.

   We can enter the transfer time in the Route Time textbox. The destination type will be chosen based on the specific target station (Destination Type: Station) or the next station in the sequence (Destination Type: Sequence).

9. *Leave flowchart module.* This module (Figures 10.18 and 10.19) seizes a resource and routes it to its destination. We can also enter the input load time and other data here.

   As shown in the figures, we can enter the load time, the transporter data or the resource data, and so on.

10. *Enter flowchart module.* This module (Figures 10.20 and 10.21) works as a station and unloads entities at stations.

    As can be seen from the figures, the Delay attribute enters an unload time. We have also chosen Release Resource in the "Transfer In" logic.

10.5.1.2   Data Modules

1. *Sequence data module.* This module consists of an ordered list of stations that an entity can visit.
2. *Distance data module.* This module consists of a set of distances between pairs of stations.
3. *Transporter data module.* This module defines a transporter, along with its load capacity and speed.

FIGURE 10.18   Leave module.

FIGURE 10.19   Leave module window.

FIGURE 10.20   Enter module.

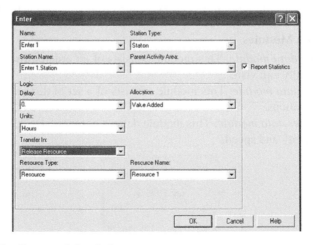

FIGURE 10.21   Enter module window.

Up to now, we have given a brief introduction to the modules that will be used in our model. Next, we give a detailed discussion of our model with the results. For convenience, we have built four models—a bidirectional model for loading activity, a bidirectional model for unloading activity, a unidirectional model for loading activity, and a unidirectional model for unloading activity.

10.6 CONCLUSION

In this chapter, the various application aspects of manufacturing simulation along with the various software applications used in the existing literature have been provided. The methodology of selecting simulation software to be purchased by an organization has also been depicted in Section 10.3.8. The list of software applications that has been shown here is not an exhaustive list. New software applications are being proposed by researchers all over the world using the latest technologies. For example, the application of cloud computing technology is being observed to be applied in various research studies. However, the recent future is expected to see large applications and simulation software with these newer technologies.

REFERENCES

1. Smith, J. S. (2003). Survey on the use of simulation for manufacturing system design and operation. *Journal of Manufacturing Systems* 22(2): 157–171.
2. Pidd, M. (1989). Choosing discrete simulation software. *OR Insight* 2(3): 6–8.
3. Hlupic, V. and Paul, R. J. (1996). Methodological approach to manufacturing simulation software selection. *Computer Integrated Manufacturing Systems* 9(1): 49–55.

11 Manufacturing and Supply Chain Simulation Packages

11.1 INTRODUCTION

This book presents various aspects of simulation. The main emphasis has been on manufacturing and supply chains. Simulation has been widely used, especially in the manufacturing sector, most likely due to the fact that manufacturing activities lie at the heart of all business activities. Thus, the manufacturing sector receives the main attention of both practitioners and researchers around the world. Supply chains are a relatively newer concept than manufacturing concepts. Manufacturing concepts are very traditional concepts, whereas supply chains have been realized for a long time, although not as long as manufacturing. We all know that supply chains are an integrated collection of business organizations that are linked together by any kind of business activities. Examples of such business organizations may be suppliers, manufacturers, distributors, wholesalers, retailers, and customers. These business organizations were always present in the business world, but the way or view of imagining them as a collection of business organizations connected in the form of a chain is quite different and known as a supply chain. The concept of a supply chain is so demanding because business organizations today are all interdependent on one another and no business organization can expect to survive without cooperation and collaboration with the neighboring echelons, in this rapidly changing dynamically competitive business world.

In this chapter, a brief discussion of some of the commonly used software applications for manufacturing and supply chains is provided. There are a plenty of manufacturing software applications available in the market. In this chapter, a brief introduction to some of the general-purpose software applications such as C and C++ is provided because these basic software applications have been used by researchers and practitioners for any basic purpose or for simulation study. A lot of proprietary software applications has also been developed from these software applications. For supply chains, too, the above-mentioned general-purpose software applications have been used in many research studies as evident from the existing literature. Thus, a brief introduction to these software applications is required.

Besides these basic general-purpose software applications, some other proprietary software applications are also discussed in this chapter. Some of these are AweSim, Arena, and a beer distribution game simulation software. This chapter is organized as follows: Section 11.2 provides a brief description of the C language.

Section 11.3 presents a brief introduction of the C++ language. Section 11.4 describes one of the traditional manufacturing software applications known as AweSim simulation software. Section 11.5 overviews a famous supply chain software known as the beer distribution game simulation proposed by the Massachusetts Institute of Technology (MIT). Section 11.6 provides the conclusion for this chapter.

11.2 INTRODUCTION TO C LANGUAGE

C language is a procedural language, which means that every step of an operation must be described in detail. It was developed in AT&T Bell Laboratories, Murray Hill, New Jersey, in 1972. It was designed and written by Dr. Denis M. Ritchie. At the time it was developed, the languages that were rather popular included FORTRAN, PL/I, ALGOL, and other similar languages. But it is a very easy-to-use, simple, and reliable language, and thus it became popular very easily. Later, many supersets of this language such as C++ and C# were developed, but C language has become a compulsory language, especially because, in many cases, it is used as a preliminary language to develop programming sense and skill.

It can be thought of as a basic language used to speak with a computer. The alphabets, digits, and symbols are used to generate constants, variables, and keywords to speak with a computer. The constants, variables, and keywords are used to develop instructions, and a set of instructions forms a program. Any program written in C involves some sort of calculation in any form. The memory cells are used to store the components of calculations, intermediate results, and final results. These memory cells are assigned names. These names are the names of variables, constants, and keywords. For example, the assignment or initialization $x = 3$ assigns the name x to a memory cell and puts the value 3 in that cell. Thus, x is called a variable.

In C, there are two types of constants—primary constants and secondary constants. Primary constants are basic integers, real, and character constants. Secondary constants are composed of primary constants. Examples of secondary constants include arrays, structures, pointers, enum, union, and so on. The keywords in C are explained by the compiler of C. Thus, these words are the reserved words. Examples of keywords include auto, if, do, for, while, int, float, char, const, switch, case, break, continue, and so on.

There are three types of instructions in C: declaration instructions, arithmetic instructions, and control instructions. Declaration instructions are responsible for declaring the variables in C. Arithmetic instructions are used to define arithmetic expressions. Control instructions are used to control the sequence of executions of the statements in C. There are four types of control structures in C: (1) sequential control structures, in which the statements are executed in sequence as they are written in the program; (2) decision control structures, in which the statements are executed based on some decisions taken; (3) loop control structures, in which a single statement or a set of statements is executed repeatedly until a certain condition fails; and (4) case control structures, in which a set of statements is executed based on a particularly selected option.

To perform the arithmetic operations properly, preferences of all the arithmetic and related operations are already defined by the C compiler. The rules that define the preferences are known as associativity rules. The arithmetic operations are performed based on the associativity rules.

The decision control structure of C mainly constitutes if–else statements. The format of if–else statements is given as follows:

```
if (specified condition is true)
        <execute set 1 of statements>;
else
        <execute set 2 of statements>;
else
.
.
.
```

The keyword "if" defines the decision control structure. A set of decisions will be taken if a particular condition is satisfied. If the condition fails, then either no decision will be taken or a different set of decisions will be taken. The conditions in C are defined in terms of logical expressions whose evaluation will result in either 0 or 1. The logical conditions include "equal to," "not equal to," "greater than," "less than," "greater than or equal to," "less than or equal to," "logical AND," "logical OR," and "exclusive OR."

C also has other types of decision structures. Some of them are switch-case statements and goto statements. The basic structure of a switch–case statement is given as follows:

```
switch (integer expression)
{
    case value 1:
                <set of statements>;
                break;
    case value 2:
                <set of statements>;
                break;
    .
    .
    .
}
```

The "value" here indicates the value of the integer expression. The "break" statement takes the control out of the switch statement after executing the set of statements under a particular "case." Another decision structure is the "goto" statement. The relevant statements are all labeled before using the goto statement. The structure of the goto statement is given as follows:

```
goto <label 1>;
```

When the control comes to this statement, the program control jumps to the statement labeled *l*. Thus, the goto statement breaks the serial execution of the set of statements forcefully, and for this reason goto is not a frequently used statement.

There are various loop structures in C language. Such loop structures include for, do–while, and while. The set of statements under a while loop get executed if the condition of the while loop is satisfied. The general structure of a while loop is given as follows:

```
while (specified condition)
{
    <set of statements>;
    <modification of variable>;
}
```

From this loop structure, it is clear that the execution of the while loop starts when the condition of the while loop is satisfied. If the condition is satisfied, then a set of statements is executed. After execution of the set of statements, the variable in the condition is modified and the control goes back to the initial condition-checking part of the loop and the loop execution starts again. The loop continues to execute until the specified condition fails. If the variable of the condition is not modified, then the loop runs infinitely until the user stops the execution of the program manually.

A special kind of while loop is called a do–while loop, where the set of statements gets executed at least once even if the condition is false. The structure of a do–while loop is given as follows:

```
do
{
    <set of statements>;
    <modification of variable>;
}while (specified condition);
```

From this loop structure, it is clear that the set of statements will be executed first followed by condition checking. Even if the evaluation of the "specified condition" results in a "false," then the set of statements still will be executed once. Thus, the purpose of such a loop is to set some compulsory decisions that will have to be executed at least once.

The most popular loop structure in C is perhaps the "for" loop. The structure of a for loop is given as follows:

```
for (initialization of variable; condition statement;
modification statement)
{
    <set of statements>;
}
```

This loop structure clearly shows that the initialization of variables, the conditional statement, and the statement for modification of variable are all kept within a single set of parentheses. So the initialization statement will be executed once at the beginning of the loop execution. After that, each time, the condition will be evaluated, and after executing the set of statements, the modification of the variable will be done. The sequence of execution (modification statement → condition statement → set of statements inside loop) will continue until the condition fails to satisfy. Several for statements, while statements, and do–while statements can be nested depending on the logic of the program. In addition, any of these loops can be executed under any of the other loops.

C language is very simple and easy to use. It has several other structures such as "functions." A function contains a set of statements to execute. A function in a C program is used when the set of statements in a function needs repeated execution. Instead of writing the same code every time, the required particular function containing the set of statements is called from the main program.

C language also provides the facility of accessing the memory locations directly through the use of "pointer" variables, which are special kinds of variables that contain the addresses of the other variables. The other structures of C include the use of macros, arrays, structures, and file input/output operations. This language can be very effectively used in system programming at the fundamental level—Windows programming, hardware programming, and graphics programming—and for interaction with simulation languages. The constructs of this language can be used for controlling different variables in a simulation package. In addition, the language itself has been used in several research studies to simulate various systems, and even the graphical utility of the language has been used to implement animated simulation. For a detailed understanding of C, the book of Yashavant [1] can be consulted.

11.3   INTRODUCTION TO C++ LANGUAGE

C++ is a type of object-oriented programming language. In an object-oriented programming language, all the concepts are represented in terms of objects. Thus, the object is the fundamental concept of C++. The required variables are mainly defined in a class, and the associated activities of a class are also defined in the class. An object is an instance of a class. Thus, the variables and functions defined in a class are all used by an object derived from that class.

The simple form of C++ is said to be a superset of C language. As an object-oriented programming language, the main characteristics of C++ include data abstraction, data encapsulation, data hiding, reusability, inheritance, and polymorphism. Among these, the term "data abstraction" refers to simplification of data. This means that instead of writing a code for each operation (as is done in C), the user can simply use the object that contains both the data and the functions to work on data. Other structures such as statements, functions, arrays, and loop structures are all similar to those of C language.

The characteristic "data encapsulation" means that the data and the functions that operate on data are both encapsulated in a class, and these encapsulated constructs can be used by objects as a whole. Only the object derived from the class can

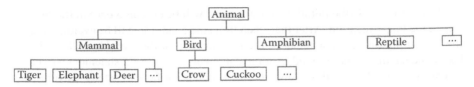

FIGURE 11.1 Example to explain inheritance.

access these data and functions, and no other object can access these data, which provides security to the existing data encapsulated in a class. Thus, data are hidden in a class and are secured. This property is known as "data hiding." Because of the compact encapsulated form, such classes can be reused in several programs. This property is called the "reusability" property. The property of "inheritance" means that the data and functions can be inherited by the child class from the parent class. For example, consider the example shown in Figure 11.1.

In the figure, animals have certain properties. For example, all animals have DNA structures, whereas all plants have RNA structures. Thus, this property is inherited by all types of animals shown in the next level—mammal, bird, amphibian, reptile and so on. Again the characteristics of mammals are all inherited by tigers, elephants, deer, and so on.

Polymorphism indicates having different forms. However, the language C++ is characterized by constructor and destructor functions, operator overloading, function overloading, inheritance, friend functions, inline functions, and so on. The inheritance property is realized by derived classes, constructors and destructors of the derived class, single- and multilevel inheritance, nested classes, virtual base classes, and virtual functions.

The literals of C++ are the same as those of the C language. However, in C++, classes are also declared and defined. Objects are created from the classes, and the class name is used like the type of the objects. Classes are defined in the same way as structures. Thus, classes can be thought of as structures along with the functions. C++ is used mostly in object-oriented programming. Thus, traditional agent-based modeling and simulations are all based on such simple object-oriented modeling. Many of the research studies from the past on agent-based modeling have been found to be based on object-oriented modeling with C++. Figure 11.2 is a simple example of how programs in C++ look.

11.4 INTRODUCTION TO AWESIM SIMULATION SOFTWARE

AweSim is a Windows-based general-purpose simulation software. It uses Visual SLAM simulation language to build its simulation networks, discrete events, and continuous event simulation models. The network model built in AweSim required no programming, although it allows developing and inserting code in Visual Basic and C. This means that AweSim can interact with Visual Basic and C. The network model developed in AweSim can be combined with discrete or continuous models developed in Visual Basic, C, or C++. AweSim is usually applied to develop

```cpp
class vehicle {
protected:
   int wheels;
   float weight;
public:
   void initialize(int in_wheels, float in_weight);
   int get_wheels(void) {return wheels;}
   float get_weight(void) {return weight;}
   float wheel_loading(void) {return weight/wheels;}
};

class car : public vehicle {
   int passenger_load;
public:
   void initialize(int in_wheels, float in_weight, int people = 4);
   int passengers(void) {return passenger_load;}
};

class truck : public vehicle {
   int passenger_load;
   float payload;
public:
   void init_truck(int how_many = 2, float max_load = 24000.0);
   float efficiency(void);
   int passengers(void) {return passenger_load;}
};

void main()
{
vehicle unicycle;

   unicycle.initialize(1, 12.5);
   cout << "The unicycle has " <<
                        unicycle.get_wheels() << " wheel.\n";
   cout << "The unicycle's wheel loading is " <<
           unicycle.wheel_loading() << " pounds on the single tire.\n";
   cout << "The unicycle weighs " <<
                        unicycle.get_weight() << " pounds.\n\n";

car sedan;

   sedan.initialize(4, 3500.0, 5);
   cout << "The sedan carries " << sedan.passengers() <<
                                        " passengers.\n";
   cout << "The sedan weighs " << sedan.get_weight() << " pounds.\n";
   cout << "The sedan's wheel loading is " <<
                     sedan.wheel_loading() << " pounds per tire.\n\n";

truck semi;

   semi.initialize(18, 12500.0);
   semi.init_truck(1, 33675.0);
   cout << "The semi weighs " << semi.get_weight() << " pounds.\n";
   cout << "The semi's efficiency is " <<
                        100.0 * semi.efficiency() << " percent.\n";
}
```

FIGURE 11.2   An example of C++ program.

```
void vehicle::initialize(int in_wheels, float in_weight)
{
    wheels = in_wheels;
    weight = in_weight;
}
void car::initialize(int in_wheels, float in_weight, int people)
{
    passenger_load = people;
    wheels = in_wheels;
    weight = in_weight;
}
void truck::init_truck(int how_many, float max_load)
{
    passenger_load = how_many;
    payload = max_load;
}
float truck::efficiency(void)
{
    return payload / (payload + weight);
}
```

FIGURE 11.2 (Continued)

models in the areas of manufacturing, transportation, military operations, health care systems, and so on.

AweSim provides a network builder to build network diagrams, forms-based builders to develop simulation run lengths, queue ranking rules, output options, text builders for user documentation, and user insert builders. The architecture of AweSim is shown in Figure 11.3. This figure shows that the heart of AweSim simulation software is the Visual SLAM simulation language. The database and text components are naturally dependent on user interaction. Various spreadsheet and text data along with various symbols are input to the Visual SLAM network model. Based on the user's input, spreadsheet data and text input and output are generated. An interactive execution environment is also responsible for developing the model. Various building blocks of AweSim used in the network diagram are shown in Figure 11.4. A sample network diagram drawn in AweSim is shown in Figure 11.5.

11.5 INTRODUCTION TO BEER DISTRIBUTION GAME SIMULATION

The beer distribution game is a simulation tool for demonstrating various aspects of supply chains. This simulation game was proposed in 1960 by MIT Sloan School of Management. The minimum number of players in the game is four. After the game is played for some time, the results are taken and analyzed.

The beer distribution game is mainly played for the behavioral study of supply chains. Most of the time the purpose is to understand the dynamics of distribution among various echelons of a supply chain for a single item. Since only the

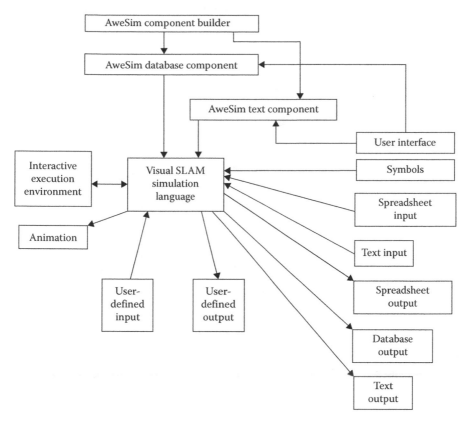

FIGURE 11.3 Architecture of AweSim.

FIGURE 11.4 Blocks used in network diagrams of AweSim.

retailer in a supply chain can see the actual customer demand, the simulation in the beer distribution game also assumes that there is only one player who can view the inventory of the end customer. All the other echelons can view only the neighboring echelon's inventory. Since verbal communication among the players is prohibited, the players suffer from disappointment, frustration, and confusion. As a result, the players try to find out the reason for any problem occurring in the system. Such a simulation can easily show the bullwhip effect in a supply chain. The bullwhip effect can be defined as the variability of order upward in a supply chain. It is one of the prime performance measures for a supply chain. Thus, the value of the bullwhip effect can reflect the performance of a supply chain. Figure 11.6 shows the typical setup of a beer distribution game simulation [2].

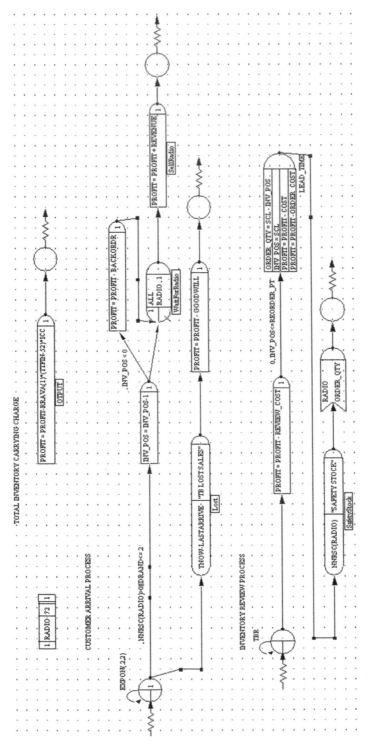

FIGURE 11.5 A sample network diagram in AweSim.

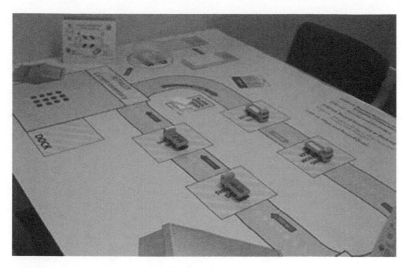

FIGURE 11.6   Typical setup of a beer distribution game.

11.6 CONCLUSION

In this chapter, two proprietary and two general-purpose software applications that are used in various manufacturing and supply chain system simulation models have been discussed. The two proprietary software applications are AweSim simulation software for manufacturing systems and the beer distribution game simulation software for supply chains. The two other general-purpose simulation software applications are C and C++. C language is a procedural language, whereas C++ is an object-oriented software. Effort has been exerted to make this chapter a short tutorial so that the reader can have urge to delve into the concepts and tools further.

REFERENCES

1. Kanetkar, Y. P. (2004). *Let Us C*. 5th Edition. New Delhi, India: BPB Publications.
2. http://supplychain.mit.edu/games/beer-game.

12 Supply Chain Simulation

12.1 INTRODUCTION

Simulation, in layman's language, is the imitation of a system to analyze the overall working of the system so as to find flaws in the system, if any, and rectify them accordingly. In the case of supply chains, such a system is difficult to develop because of the complexity of the system. A supply chain can be thought of as an integrated collection of interrelated business organizations that are linked by business transactions. Thus, a supply chain may consist of suppliers, manufacturers, distributors, wholesalers, and retailers (Figure 12.1). But the structure of a real supply chain is not so simple and not serial in nature. Rather than a simple chain of serially connected echelons (business organizations), a supply chain is actually a network of business organizations that engage in complex transactions with each other.

For example, a wholesaler or a big retailer may directly purchase from a manufacturer or there may exist a network or distributor organization. Even in the case of third-party logistics (3PL), often we see a network of companies working together, each being a customer of the other. Moreover, a local warehouse may be controlled by a central warehouse. There are numerous other similar examples of the complex transactions leading to the complex structure of a supply chain.

The major drivers of a supply chain are as follows:

1. *Production.* It actually concerns the factories. Factories can be built on the basis of any of the following two approaches:
   a. Product focus—A range of different operations are performed to make a given product line from the fabrication of different product parts to the assembly of these parts.
   b. Functional focus—This approach concentrates on performing some operations such as only making a select group of parts or only doing assembly.
2. *Inventory.* It first considers the approaches in warehousing. There are three main approaches in warehousing:
   a. Stock keeping unit (SKU) storage—This is a traditional approach where all of a given type of product are stored together. This is an efficient and easy way to store products.
   b. Job lot storage—In this approach, all the different products related to the needs of a certain type of customer or the needs of a particular job are stored together. This allows for an efficient picking and packing operation but usually requires more storage space than the traditional SKU storage approach.

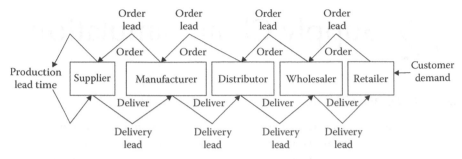

FIGURE 12.1   A serial supply chain.

   c. Cross-docking—An approach that was pioneered by Wal-Mart in its
      drive to increase efficiency in its supply chain. In this approach, the
      product is not actually warehoused in the facility. Instead, the facility
      is used to house a process in which trucks from suppliers arrive and
      unload large quantities of different products.
      There are three basic decisions to make regarding the creation and
      holding of inventory:
      a. Cycle inventory
      b. Safety inventory
      c. Seasonal inventory
3. *Location and distribution.* It is a vast topic, and numerous aspects of these
   two factors can be discussed. Distribution strategy deals mainly with
   distribution strategies, distribution channels, modes of distribution, and so
   on. Location decisions deal with deciding on the location of various echelons of a supply chain.
4. *Transportation.* There are six basic modes of transport that a company can
   choose from. They are as follows:
   a. Ship
   b. Rail
   c. Pipelines
   d. Trucks
   e. Airplanes
   f. Electronic transport
   The pipeline mode is restricted to commodities that are liquids or gases
   such as water, oil, and natural gas.
5. *Information.* It is used for two purposes in any supply chain:
   a. Coordinating daily activities
   b. Forecasting and planning

Due to the complexity of practical supply chains, the simulation of supply chains is
a difficult task. In addition, the integration of the various organizations of a supply
chain also implies the coordination among them, which is also a difficult task because
of various barriers from the organizational point of view. However, different researchers from all over the world have tried different ways to solve the various problems of

supply chains, and to accomplish their purpose, they have simulated some functions of supply chains. Thus, the various aspects of supply chains will have to be considered before we delve deeper into the concept. The various areas where simulation studies for supply chains have been conducted are depicted in Section 12.2.

The remaining sections of this chapter are arranged as follows: Section 12.2 mentions the various areas of supply chains where different types of simulation studies have been applied. Section 12.3 depicts different types of supply chain simulations as evident from the existing literature and gives a particular emphasis on parallel distributed simulations for supply chains. Section 12.4 mentions the various kinds of software packages used in the literature for simulating supply chain applications and also describes some dominant simulation software packages and tools for supply chain simulations as evident from the existing literature. Section 12.5 provides the conclusion for this chapter.

12.2 AREAS OF SUPPLY CHAIN SIMULATION

The area of supply chains is so vast that it actually encompasses all business organizations as well as all business activities. Therefore, we will discuss only a limited number of areas of applications of supply chains. By definition, the supply chain also incorporates the manufacturing functions. Thus, all the areas of manufacturing can also be indirectly referred as the areas of supply chains.

However, the existing literature has emphasized some of the areas, particularly the specific areas of supply chains. Some of the significant ones among such areas are as follows:

- Supply chain network design
- Supply chain strategic decision support system
- Demand and sales planning
- Supply chain planning
- Inventory planning
- Logistics management
- Supply chain scheduling
- Supply chain event scheduling
- Distribution and transport planning
- Reduction planning and scheduling
- Supply chain performance measure
- Supply chain coordination

A significant number of research studies have been published in each of the above-mentioned areas. However, as mentioned earlier, the areas are not limited to this list only. Numerous aspects and areas can be incorporated as the area of supply chains. In addition, each of the above-listed areas has subareas. For example, the supply chain's performance measuring system may depend on the numerous performance factors to be considered. These factors depend on the type of supply chain under consideration, such as product supply chain, service supply chain, electronic supply chain, automobile supply chain, green supply chain, and vegetable supply chain. Some of these areas are briefly described in Sections 12.2.1 through 12.2.4.

12.2.1 DISTRIBUTION IN SUPPLY CHAIN

There are various types of distributions in supply chains. These are given as follows:

1. *Direct shipment distribution strategy.* The manufacturer or supplier delivers goods directly to retail stores. The advantages of this strategy are as follows:
   a. The retailer avoids the expenses of operating a distribution center.
   b. Lead times are reduced.

   The disadvantages of this strategy are as follows:
   a. There is no central warehouse.
   b. The manufacturer and distributor transportation costs increase because smaller trucks must be sent to more locations.
2. *Intermediate inventory storage point strategy.* Inventory is stored at an intermediate point. The various strategies under this category are as follows:
   a. Traditional warehousing
   b. Cross-docking
   c. Inventory pooling
3. *Strategic alliance (3PL).* In this strategy, there are three parties involved— manufacturer, supplier, and third party (performing all the parts of the firm's materials management and product distribution functions).

   The advantages of this strategy are as follows:
   a. Firms can focus on core strengths.
   b. The better 3PL providers constantly update their information technology and equipment.
   c. 3PL may also provide greater flexibility for a company in terms of geographic locations since suppliers may require paid replenishment, which in turn may require regional warehousing.

12.2.2 COLLABORATIVE PLANNING, FORECASTING, AND REPLENISHMENT

The collaborative planning, forecasting, and replenishment (CPFR) process is divided into three activities: planning, forecasting, and replenishment. Within each activity, there are several steps. These steps are depicted under each of the categories as follows:

Collaborative planning
- Negotiate a front-end agreement that defines the responsibilities of the companies that will collaborate with each other.
- Build a joint business plan that shows how the companies will work together to meet demand.

Collaborative forecasting
- Create sales forecasts for all the collaborating companies.
- Identify any exceptions or differences between companies.
- Resolve the exceptions to provide a common sales forecast.

Collaborative replenishment
- Create order forecasts for all the collaborating companies.
- Identify any exceptions between companies.

- Resolve the exceptions to provide an efficient production and delivery schedule.
- Generate actual orders to meet customer demand.

12.2.3 SUPPLY CHAIN PERFORMANCE MEASURE

The area of supply chain performance measure has attracted considerable attention from both researchers and industrial practitioners over the past decade. A significant number of research studies related to the performance measure of supply chains are observed in the existing literature. Competition in today's business world is between supply chains and not between companies. Thus, from this perspective, measuring supply chains and the related research are essential for the survival of business organizations in today's highly dynamic and competitive business environment. In addition, supply chain performance measure is also relevant from the perspective of globalization, increased outsourcing, increasing demands of customers, shorter product life cycles, increasing agility, increasing collaboration among the organizations, and internal and external integration of various business activities.

The performance measure is essential since measurement always improves a system, accomplishing anything measured becomes possible and easier to achieve, and effective management is possible based on measurement. Through performance measurement, an organization can assess whether it has improved or degraded over time. According to Gunasekaran and Kobu [1] and Akyuz and Erkan [2], the purposes of a performance measurement system are as follows:

1. Satisfy customer needs
2. Identify success
3. Minimize waste
4. Improve the decision-making process
5. Study the progress of an organization
6. Identify bottlenecks
7. Understand the system as a whole
8. Collaborate and cooperate to a greater extent
9. Improve communication

However, the measurement of performance depends on the type of supply chain being studied. Thus, the performance measure system will differ in electronic supply chains, green supply chains, product supply chains, or service supply chains. The reason for the difference lies in the difference of the type of final products in these supply chains. In green supply chains, the final products are more perishable in nature than the other supply chains, whereas the final products of electronic supply chains are mostly intangible or soft products that may become obsolete over time. Waste minimization is the prime concern for vegetable supply chains, whereas obsolescence is the prime concern for electronic supply chains.

The most widely used overall quantitative performance measures as evident from the existing literature are as follows: (1) total cost, (2) total inventory, (3) service level, and (4) bullwhip effect. The term "bullwhip" was originally coined by the logistics executives at Procter & Gamble (P&G). The bullwhip effect can be defined as the variability of orders upward in a supply chain. The above-mentioned performance measures are the quantitative measures of the performance of a supply chain.

The existing literature investigates various performance measurement models. The most significant among them are mentioned in the research study of Estampe et al. [3], who discuss a total of 16 benchmark measurement models for supply chain performance as shown in Figure 12.2. These are briefly discussed as follows:

1. *Activity-based costing (ABC).* It focuses on the cost and margin to measure supply chain performance. Tsai and Hung [4] apply goal programming to integrate ABC with supply chain performance for a green supply chain. The authors also use an analytic hierarchy process (AHP) to arrive at the final conclusion and show that the proposed approach can be used for further improvement of green supply chains. LaLonde and Terrance [5] discussed the direct product profitability (DPP) as another costing technique other than ABC and proved that ABC is the better costing technique. Dekker and Van Goor [6] showed the use of ABC in supply chain decision making through a case study of a pharmaceutical organization.

   Some of the other significant research studies dealing with ABC for supply chain performance include the studies by Binshan et al. [7] and Bagchi et al. [8].

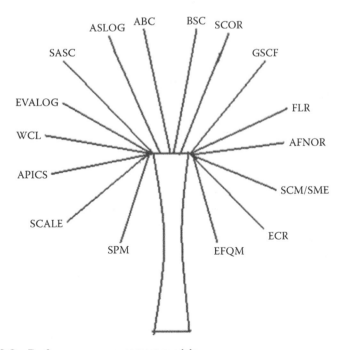

FIGURE 12.2   Performance measurement models.

2. *Balanced scorecard (BSC)*. It performs balanced measures based on the company's strategy from the aspects of customer, finance, processes, and innovation growth. Human decision making is taken into account while measuring supply chain performance through BSC. The relevant research studies include the studies by Brewer and Speh [9], Bhagwat and Sharma [10], Hans-Jörg et al. [11], and Hervani et al. [12].

3. *Supply chain operation reference (SCOR)*. It has been used to measure the performance of supply chains based on the decision making at various levels of management: strategic, tactical, and operational. SCOR measures four dimensions of performance, namely, cost, turnover, reliability, and flexibility. The relevant research studies in this direction include the studies by Lockamy and McCormack [13], Hans-Jörg et al. [11], Chan and Qi [14], and Bolstorff and Rosenbaum [15].

4. *Global supply chain framework (GSCF)*. It is based on seven-dimensional processes for strategic, tactical, and operational levels of management. The seven dimensions are customer relationship management (CRM), customer service management (CSM), demand management, order fulfillment, manufacturing flow management, supplier relationship management, product development, and commercialization returns management. Both SCOR and GSCF focus on the link between the internal process of a company and the structure of a supply chain and its performance. The significant research studies on GSCF for supply chain performance measurement include the studies by Chan and Qi [16], Lambert and Cooper [17], Ellram et al. [18], and Bendavid et al. [19].

5. *ASLOG audit*. It focuses on the cross-functional processes in a supply chain to measure the benefits and negative aspects of logistics procedures. It is mainly applicable to companies with low or medium maturity. Some of the research studies on ASLOG audits include the studies by Gruat La Forme et al. [20], Estampe et al. [3], Halley and Guilhon [21], and Wattky and Neubert [22,23].

6. *Strategic audit supply chain (SASC)*. It takes processes, organization, and information technology for an organization to assess the performance of logistics. There are a limited number of research studies in this direction, for example, the study by Hadiguna et al. [24].

The above-mentioned models are benchmark models. In addition, there are other models such as the framework for logistics research (FLR), Global EVALOG, world-class logistics (WCL), AFNOR FD X50-605, SCM/SME, APICS, efficient customer response (ECR), EFQM, supply chain advisor level evaluation (SCALE), and strategic profit model (SPM) [3]. Research studies in these models are very few and thus cannot be regarded as benchmark models. However, the existing literature shows SCOR, BSC, and ABC as the most widely used models for measuring supply chain performance.

12.2.3.1 Various Performance Measures and Metrics

Figure 12.3 shows the approaches for various performance measures and metrics as evident from the existing literature.

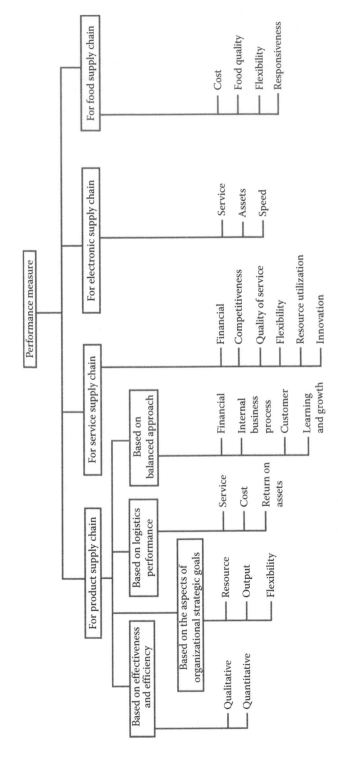

FIGURE 12.3 Performance measures and metrics.

The performance measure depends on the nature of the final product in a supply chain as evident from the existing literature. Thus, Figure 12.2 has differentiated the metrics for different types of supply chains based on their final product, product supply chain, service supply chain, electronic supply chain, and food supply chain. Although the food supply chain is a part of a product supply chain, the products of a food supply chain are perishable, whereas the final products of a product supply chain are either nonperishable or much less perishable.

The performance measure for product supply chains is further subdivided based on the various aspects delivered by the existing literature. Although there are other divisions, these are observed to be the widely used divisions as evident from the existing literature. Thus, the measurement can be based on efficiency and effectiveness [25], the aspects of organizational strategies and goals [26], logistics performance [9], and the balanced approach [10,27]. In addition, there are separate measurements for service supply chains [28], electronic supply chains [29], and food supply chains [30]. The balanced approach is mainly based on SCOR and BSC models that are widely used in the existing literature. The metrics provided by these measurement methods are given briefly as follows:

1. Measurement based on efficiency and effectiveness [25]
   a. Qualitative measures for which no numerical measurement is possible. These are as follows:
      i. Customer satisfaction that is further classified into pretransaction satisfaction, during-transaction satisfaction, and posttransaction satisfaction
      ii. Flexibility in responding to demand fluctuation
      iii. Integration between material and information communication
      iv. Risk
      v. Consistency of supplier delivery
   b. Quantitative measures that can be measured numerically. These are as follows:
      i. Based on the cost
         A. Total cost of entire supply chain
         B. Quantity of product sold
         C. Total profit (revenue minus cost)
         D. Total inventory cost
         E. Return on investment (net profit divided by capital)
      ii. Based on customer responsiveness
         A. Fill rate (fraction of customer orders filled)
         B. Product lateness (promised product delivery date minus actual delivery date)
         C. Customer response time (time between placing an order and receiving order by customer)
         D. Manufacturing lead time (time to manufacture a product)
         E. Function duplication (number of same business functions performed by more than one entity)

2. Measurement based on the aspects of organizational strategies and goals [26]
  a. Resource measure
      i. Inventory level
      ii. Personnel required
      iii. Equipment utilization
      iv. Usage of energy
      v. Total cost of resources
      vi. Distribution cost
      vii. Manufacturing cost
      viii. Return on investment
  b. Output measure
      i. Customer response (qualitative measure)
      ii. Product quality (qualitative measure)
      iii. Number of final products produced
      iv. Time required to produce a final product
      v. On-time deliveries
          A. Product lateness
          B. Average lateness
          C. Average earliness of orders
          D. Percentage of on-time deliveries
      vi. Sales (total revenue)
      vii. Total profit (revenue minus cost)
      viii. Fill rate (fraction of customer orders filled)
          A. Target fill rate
          B. Average fill rate
      ix. Backorder/stockout
          A. Stockout probability
          B. Number of backorders
          C. Number of stockouts
          D. Average backorder level
      x. Customer response time
      xi. Manufacturing lead time
      xii. Shipping errors
  c. Flexibility measure
      i. Ability to adjust to volume change
      ii. Ability to adjust to fluctuation in supplier schedules
      iii. Ability to adjust to change in customer demands
      iv. Ability to adjust to manufacturing schedule changes
3. Measurement based on logistics performance [9]
  a. Service measure
      i. Order cycle time
      ii. Fill rate
      iii. Damage rate
      iv. Error rate in picking orders
      v. Achievement of "perfect order"

    b. Cost measure
       i. Ordering cost
       ii. Logistics cost
       iii. Storage cost
       iv. Holding cost
    c. Return of assets measure
       i. Return on investment
       ii. Net earning
4. Measurement based on the balanced approach [10,27]
    a. Performance for planned orders
       i. Order entry method
       ii. Order lead time
       iii. Customer order path (different route and non-value-adding services)
    b. Extent of partnership
       i. Level and degree of information sharing
       ii. Buyer–vendor cost-saving initiatives
       iii. Extent of mutual cooperation
       iv. Extent of supplier involvement
       v. Extent of mutual efforts in problem solving
    c. Customer service and satisfaction
       i. Flexibility to accommodate demand fluctuation
       ii. Customer query time
       iii. Posttransaction measures of customer service
          A. Service level compared to competitors
          B. Customer perception of service
    d. Product-level measures
       i. Range of products or services
       ii. Capacity utilization
       iii. Effectiveness of scheduling techniques
    e. Delivery-related metrics
       i. Delivery-to-request date
       ii. Delivery-to-commit date
       iii. Order fill lead time
       iv. Number of faultless notes invoiced
    f. Finance- and logistics-related metrics
       i. Cost of assets
       ii. Return on investment
       iii. Total inventory cost
5. Measurement for service supply chains [28]
    a. Financial
       i. Asset turnover (indicating profitability)
       ii. Labor and capital cost (indicating liquidity)
       iii. Profit per service (indicating capital structure)
    b. Competitiveness
       i. Ability to win new customers
       ii. Customer loyalty

    c. Quality of service
       i. Relationship between customers and organization
      ii. Customer satisfaction
    d. Flexibility
       i. Flexibility to increase volume
      ii. Flexibility to change delivery speed
     iii. Flexibility to change specifications in service design
     iv. Flexibility to change capacity strategies
      v. Flexibility to adjust part-time and floating staff
     vi. Flexibility to change demand
    vii. Flexibility to price and promotion strategies
    e. Resource utilization
       i. Utilization of facilities
      ii. Utilization of equipments
     iii. Utilization of staff
    f. Innovation
       i. Measurement for success of innovation
      ii. Extent of innovation
6. Measurement for electronic supply chains [29]
    a. Service metrics
       i. For build-to-stock situations
         A. Line item fill rate
         B. Complete order fill rate
         C. Delivery process on time
         D. Number of backorders
         E. Cost of backorders
         F. Aging of backorders
      ii. For build-to-order situations
         A. Quoted customer response time
         B. Percentage of on-time completion
         C. Delivery process on time
         D. Number of late orders
         E. Cost of late orders
         F. Aging of late orders
    b. Asset/inventory metrics
       i. Monetary value of inventory
      ii. Timely supply of inventory turns
    c. Speed metrics
       i. Cycle time at a node
      ii. Supply chain cycle time
     iii. Cash conversion cycle
     iv. "Upside" flexibility (flexibility to meet high-tech requirements)
7. Measurement for food supply chain [30]
    a. Efficiency
       i. Cost

          A. Production cost
          B. Distribution cost
          C. Transaction cost
      ii. Profit
     iii. Return on investment
     iv. Inventory
   b. Food quality
      i. Product quality
          A. Sensory properties and shelf life
          B. Product safety and health
          C. Product reliability and convenience
      ii. Process quality
          A. Production system
          B. Environmental aspects
          C. Marketing
   c. Flexibility
      i. Customer satisfaction
      ii. Volume flexibility
     iii. Delivery flexibility
     iv. Number of backorders
      v. Number of lost sales
   d. Responsiveness
      i. Fill rate
      ii. Product lateness
     iii. Customer response time
     iv. Lead time
      v. Shipping errors
     vi. Customer complaints

Although there are overlaps among the different views of measures, they have been listed separately because of the differences in views. In addition to these classifications, there are also other different types of views toward the types of measurements and metrics.

Olugu et al. [31] investigated the performance measures for an automobile green supply chain. They first classified the supply chains into a forward chain (starting from supplier and proceeding toward customer) and a backward chain (starting from customer), and then showed the measures under each of these two chains. The performance measures of an automobile green supply chain are as follows:

1. Forward chain measures
   a. Upstream measures—supplier commitment
      i. Level of supplier environmental certification
      ii. Level of supplier performance on sustainability
     iii. Number of supplier initiatives on environmental management
     iv. Level of disclosure of environmental initiative to the public
      v. Level of supplier preprocessing of raw materials

   b. Midstream measures
- i. Greening cost
  - A. Cost of environmental complaint
  - B. Cost of energy consumption
  - C. Cost of environment-friendly materials
  - D. Green cost per revenue
- ii. Level of process management
  - A. Availability of process optimization for waste reduction
  - B. Level of spillage, leakage, and pollution control
  - C. Level of waste generated from production
  - D. Quantity of utility used
  - E. Number of violations of environmental regulations
- iii. Product characteristics
  - A. Level of recycled products
  - B. Level of disposed products
  - C. Availability of eco-labeling
  - D. Level of biodegradable content in products
  - E. Level of usage of design-for-assembly in products
  - F. Level of market share controlled by green products
- iv. Management commitments
  - A. Level of management efforts to motivate employees
  - B. Availability of environment evaluation schemes
  - C. Availability of environmental audit systems
  - D. Availability of mission statement on sustainability
  - E. Number of environmental management initiatives
  - F. Level of management effort to enlighten customers on sustainability
  - G. Availability of environmental reward systems
  - H. Level of management effort to motivate suppliers
- v. Traditional supply chain costs
  - A. Total supply chain cost
  - B. Delivery cost
  - C. Inventory cost
  - D. Information sharing cost
  - E. Ordering cost
- vi. Responsiveness
  - A. Order lead time
  - B. Product development cycle time
  - C. Manufacturing lead time
  - D. Total supply chain cycle time
  - E. Number of on-time delivery
- vii. Quality
  - A. Customer satisfaction
  - B. Delivery unreliability
  - C. Amount of scrap and rework
  - D. Availability of green product warranty

      viii. Flexibility
- A. Flexibility to demand fluctuations
- B. Flexibility to change delivery schedule
- C. Flexibility to production schedule

  c. Downstream measures
    i. Customer perspective
- A. Level of customer interest in green products
- B. Level of customer satisfaction from green products
- C. Level of customer dissemination of green information

2. Backward chain measures
  a. Upstream measures
    i. Level of customer cooperation
    ii. Level of customer-to-customer dissemination of information
    iii. Level of understanding of green processes by customers

  b. Midstream measures
    i. Recycling cost
- A. Cost of processing of recyclables
- B. Cost of storing and segregation of recyclables
- C. Cost of disposal for hazardous and unprocessed waste

    ii. Material features
- A. Level of waste generated
- B. Ratio of materials recycled to recyclable materials
- C. Material recovery time

    iii. Management commitment
- A. Level of motivation to customers
- B. Availability of standard operating procedure for the collection of end-of-life vehicles (ELVs)
- C. Availability of collection centers
- D. Availability of waste management schemes

    iv. Recycling efficiency
- A. Percentage decrease in recycling time
- B. Availability of recycling standards
- C. Availability of standard operating procedures
- D. Percentage decrease in utility usage during recycling
- E. Efficiency of shredders and dismantlers
- F. Percentage reduction in emission and waste generated

  c. Downstream measures
    i. Extent of delivery from suppliers to manufacturers
    ii. Certification system for suppliers in the recycling process
    iii. Number of supplier initiatives in the recycling process

In addition, Christopher and Towill [32] mentioned four performance measures: quality, cost, lead time, and service level. Van der Vorst [33] proposed three levels of performance indicator for food supply chains, namely, supply chain, organization, and process. The indicators at the supply chain level are product availability, quality, responsiveness, delivery reliability, and total supply chain cost. The indicators at the organization level are inventory level, throughput time, responsiveness, delivery reliability, and total organizational cost. At the process level, the indicators are responsiveness, throughput time, process yield, and process cost. Gunasekaran et al. [34] measured performance at three levels of management—strategic level, tactical level, and operational level. Li and O'Brien [35] measured supply chain performance at two levels—chain level and operational level.

The other significant research studies in measuring supply chain performance include the studies by Talluri and Baker [36], Claro et al. [37], Gunasekaran et al. [38], Murphy et al. [39], Berry and Naim [40], Thonemann and Bradley [41], Agarwal et al. [42], Min and Zhou [43], Govindan et al. [44], Paksoy et al. [45], Correia et al. [46], Li et al. [47], Cai et al. [48], and Kroes [49].

12.2.4 METHODOLOGIES USED IN THE EXISTING LITERATURE

The research studies in supply chain performance measurement have used a variety of techniques, including AHP, exploratory factorial analysis (EFA), analytic network process (ANP), simulation, linear regression, structural equations, discriminate analysis, multicriteria analysis, and quantitative analysis. Structural analysis has been used in cases where the performance depends on one or more factors. However, all these techniques endeavor to improve the performance of supply chains.

However, based on the areas of supply chains that have been conducted, Section 12.3 gives a glimpse into the various types of simulations.

12.3 TYPES OF SUPPLY CHAIN SIMULATIONS

Depending on the type of approach followed while simulating various aspects of a supply chain, several types of supply chain simulations are observed in the existing literature. These types are discussed below.

The first among all the types is *parallel discrete event simulation*. It is nothing but the execution of a simulation program on multiple parallel processors. This type of simulation is based on a collaboration of programs. Thus, this type of execution has all the advantages of a parallel execution. Some of these advantages are as follows:

- Multiple copies of the program can be run on several machines.
- Each of the multiple copies may be for a separate purpose.
- The execution time for all these differently purposed programs is less than running those programs on only one processor.
- Various aspects of a simulation software can be tested in a short time by this process.

The second type of simulation is *distributed simulation*, where various modules of the same simulation package can be executed on various machines. This means that the entire task of simulation can be divided into subtasks and each of these subtasks can be run on a separate machine. It has the following advantages:

- The entire software modules are divided into several machines. Thus, each machine gets fewer burdens.
- Since the entire task is divided among several subtasks to be performed by several machines, the time taken to accomplish the entire task is much less.
- If a machine faces some problems in executing the module assigned to it, the other machine can take up the task later.
- The security of the entire package is now not the responsibility of any one machine.

The distributed simulation is very demanding because of the following reasons:

- It reduces the execution time.
- It can be used to reproduce graphically distributed systems.
- It can be used to integrate different simulation models, tools, and languages.
- It can be used to increase the tolerance of the failure of simulation.

The third type of simulation is *parallel distributed simulation*. This type of simulation incorporates the advantages of both the parallel simulation and the distributed simulation. Thus, the advantages of this type of simulation combine the advantages of both the parallel simulation and the distributed simulation.

Parallel and distributed simulations can be either analytical simulations or distributed virtual environments. An analytical simulation analyzes the quantitative aspects of a system, whereas a distributed virtual environment is a virtual computing environment that can be used for several purposes. Parallel and distributed simulations can be either a networked structure or a centralized structure. They can facilitate the simulation of supply chains in the following ways:

- Complex supply chain simulation can be facilitated by parallel and distributed simulations.
- It facilitates simulation of a supply chain whose echelons are distributed geographically.
- It may help in reducing the execution time of supply chain simulation.

There is also another recent trend of *simulation by using virtualization*. The concept of cloud computing, in this case, may be implemented through virtualization. Thus, various echelons may be represented by separate virtual machines and their interactions may represent the interaction among the supply chain echelons. This is an emerging simulation trend, and very hardcore computer knowledge is required to implement such ideas of simulation.

However, after deciding on the areas of simulation for supply chains, the next difficult task is to choose a way to implement them by using computers. Section 12.4 presents the types of software packages that may be used for supply chain simulations.

12.4   TYPES OF SUPPLY CHAIN SIMULATION SOFTWARE

A wide range of concepts for supply chain simulation have been used, as evident from the existing literature. The simulation concepts and software applications range from general-purpose software to supply chain simulations along with simulations in other fields of study. Such general-purpose software packages include the software packages mentioned in Chapter 10. In addition, there are other software packages for parallel distributed simulation approaches, some of which are listed as follows:

- CMB-DIST
- MPI-ASP
- GRIDS
- HLA
- DEVS/CORBA

There are separate sets of software packages for supply chain design, supply chain optimization, and so on. Table 12.1 provides examples of some significant software packages for supply chain design.

Among the several supply chain optimization software packages, the most significant ones include Xpress-MP and CPLEX. In addition, general-purpose software packages has also been used for supply chain simulations. Such types of software programs include C++, C#, .NET, and Java.

Other kinds of simulation tools that are available include (1) simulation with beer distribution game software, (2) simulation with systems dynamics, and (3) simulation with cloud computing (as a future generation tool).

The characteristics of a beer distribution game are depicted as follows:

- The beer distribution game proposed by the Massachusetts Institute of Technology (MIT), Cambridge, Massachusetts, is a widely used simulation tool for supply chains.
- It is basically a board game that is mainly played for studying the inventory behavior. It is especially useful for studying the famous bullwhip effect,

TABLE 12.1

Examples of Software Packages for Supply Chain Design

Software Names	Vendors
Supply Chain Builder	Simulation Dynamics, Inc.
Supply Chain Guru	LLamasoft
SAILS	Insight
Opti-Net	TechnoLogix Decision Sciences
LOPTIS	Optimal Software
Supply Chain Strategist	i2 Technologies
NETWORK	Supply Chain Associates
CAST	Barloworld Optimus
PRODISI SCO	Prologs

which indicates the variability of orders upward in a supply chain. The bullwhip effect is one of the prime performance measures of a supply chain. The basic rules of the game are as follows:

- Pick a team name.
- Each player fills out a record sheet—inventory/backlog and orders.
- There should be no communication between players.
- The end customer orders are not revealed by the retailers to the other echelons.
- Play the game for a specific period of time units.
- Complex inventory situations can be simulated with the beer distribution game.
- This game depicts how several decisions made by the players make an aggregate impact on the inventory and therefore on the performance.
- This game identifies the external factors that can affect the external behavior.
- This game depicts the importance of coordination among the supply chain partners.

12.5   CONCLUSION

In this chapter, an introductory overview of the concept of supply chain simulation has been provided. Various types of simulations applied in cases of supply chains have been discussed. The areas where the simulation studies have been conducted have also been identified for supply chain scenarios. The software packages that have been used by different researchers in the area of supply chain observed in the existing literature have also been mentioned in this chapter.

The future of supply chain simulation can benefit from the latest technologies. Such technologies include .NET software and especially software for virtualization that may enable the researchers to analyze various problems of supply chains.

REFERENCES

1. Gunasekaran, A. and Kobu, B. 2007. Performance measures and metrics in logistics and supply chain management: A review of recent literature (1995–2004) for research and applications. *International Journal of Production Research* 45(12): 2819–2840.
2. Akyuz, G. A. and Erkan, T. E. (2010). Supply chain performance measurement: A literature review. *International Journal of Production Research* 48(17): 5137–5155.
3. Estampe, D., Lamouri, S., Paris, J.-L., and Brahim-Djelloul, S. (2010). A framework for analysing supply chain performance evaluation models. *International Journal of Production Economics* 142(2): 247–258.
4. Tsai, W.-H. and Hung, S.-J. (2009). A fuzzy goal programming approach for green supply chain optimisation under activity-based costing and performance evaluation with a value-chain structure. *International Journal of Production Research* 47(18): 4991–5017.
5. LaLonde, B. J. and Terrance, L. P. (1996). Issues in supply chain costing. *International Journal of Logistics Management* 7(1): 1–12.
6. Dekker, H. C. and Van Goor, A. R. (2000). Supply chain management and management accounting: A case study of activity-based costing. *International Journal of Logistics Research and Applications* 3(1): 41–52.

7. Binshan, L., James, C., and Robert, K. S. (2001). Supply chain costing: An activity-based perspective. *International Journal of Physical Distribution & Logistics Management* 31(10): 702–713.
8. Bagchi, S., Buckley, S. J., Ettl, M., and Lin, G. Y. (1998). Experience using the IBM supply chain simulator. *Proceedings of the 1998 Winter Simulation Conference*, ACM 2000, Orlando, FL, December 13–16, pp. 1387–1394.
9. Brewer, P. C. and Speh, T. W. (2000). Using the balanced scorecard to measure supply chain performance. *Journal of Business Logistics* 21(1): 75–93.
10. Bhagwat, R. and Sharma, M. K. (2007). Performance measurement of supply chain management: A balanced scorecard approach. *Computers & Industrial Engineering* 53(1): 43–62.
11. Hans-Jörg, B., Michael, K., and Antonius, V. H. (2002). Analysing supply chain performance using a balanced measurement method. *International Journal of Production Research* 40(15): 3533–3543.
12. Hervani, A. A., Helms, M. M., and Sarkis, J. (2005). Performance measurement for green supply chain management. *Benchmarking: An International Journal* 12(4): 330–353.
13. Lockamy III, A. and McCormack, K. (2004). Linking SCOR planning practices to supply chain performance: An exploratory study. *International Journal of Operations & Production Management* 24(12): 1192–1218.
14. Chan, F. T. S. and Qi, H. J. (2003a). An innovative performance measurement method for supply chain management. *Supply Chain Management: An International Journal* 8(3): 209–223.
15. Bolstorff, P. and Rosenbaum, R. (2007). *Supply Chain Excellence: A Handbook for Dramatic Improvement Using the SCOR Model.* New York: American Management Association.
16. Chan, F. T. S. and Qi, H. J. (2003b). Feasibility of performance measurement system for supply chain: A process-based approach and measures. *Integrated Manufacturing Systems* 14(3): 179–190.
17. Lambert, D. M. and Cooper, M. C. (2000). Issues in supply chain management. *Industrial Marketing Management* 29(1): 65–83.
18. Ellram, L. M., Tate, W. L., and Billington, C. (2004). Understanding and managing the services supply chain. *Journal of Supply Chain Management* 40(4): 17–32.
19. Bendavid, Y., Lefebvre, É., Lefebvre, L. A., and Fosso-Wamba, S. (2009). Key performance indicators for the evaluation of RFID-enabled B-to-B e-commerce applications: The case of a five-layer supply chain. *Information Systems and e-Business Management* 7(1): 1–20.
20. Gruat La Forme, F.-A., Genoulaz, V. B., and Campagne, J.-P. (2007). A framework to analyse collaborative performance. *Computers in Industry* 58(7): 687–697.
21. Halley, A. and Guilhon, A. (1997). Logistics behaviour of small enterprises: Performance, strategy and definition. *International Journal of Physical Distribution & Logistics Management* 27(8): 475–495.
22. Wattky, A. and Neubert, G. (2004). Integrated supply chain network through process approach and collaboration. *Proceedings of the 2nd IEEE International Conference on Industrial Informatics*, IEEE, Berlin, Germany, June 24–26, pp. 58–63.
23. Wattky, A. and Neubert, G. (2005). Improving supply chain performance through business process reengineering. In *Knowledge Sharing in the Integrated Enterprise: Interoperability Strategies for the Enterprise Architect.* Bernus, P. and Fox, M. (Eds.), New York: Springer, pp. 337–349.
24. Hadiguna, R. A., Jaafar, H. S., and Mohamad, S. (2011). Performance measurement for sustainable supply chain in automotive industry: A conceptual framework. *International Journal of Value Chain Management* 5(3/4): 232–250.

25. Beamon, B. M. (1998). Supply chain design and analysis: Models and methods. *International Journal of Production Economics* 55(3): 281–294.
26. Beamon, B. M. (1999). Measuring supply chain performance. *International Journal of Operations & Production Management* 19(3): 275–292.
27. Kleijnen, J. P. C. and Smits, M. T. (2003). Performance metrics in supply chain management. *Journal of the Operational Research Society* 54(5): 507–514.
28. Cho, D. W., Lee, Y. H., Ahn, S. H., and Huang, M. K. (2012). A framework for measuring the performance of service supply chain management. *Computers & Industrial Engineering* 62(3): 801–818.
29. Hausman, W. H. (2002). Supply Chain Performance Metrics. In *The Practice of Supply Chain Management*, Billington, C., Harrison, T., Lee, H., and Neale, J. (Eds.), Dordrecht, The Netherlands: Kluwer Academic Publishers, pp. 1–15.
30. Aramyan, L. H. (2007). Measuring supply chain performance in the agri-food sector. PhD Thesis, Wageningen University, Wageningen, The Netherlands.
31. Olugu, E. U., Wong, K. Y., and Shaharoun, A. M. (2011). Development of key performance measures for the automobile green supply chain. *Resources, Conservation and Recycling* 55(6): 567–579.
32. Christopher, M. and Towill, D. R. (2001). An integrated model for the design of agile supply chains. *International Journal of Physical Distribution & Logistics Management* 31(4): 235–246.
33. Van der Vorst, J. G. A. J. (2000). Effective food supply chains. Generating, modelling and evaluation supply chain scenarios. PhD thesis, Wageningen University, Wageningen, The Netherlands.
34. Gunasekaran, A., Patel, C., and McGaughey, R. (2004). A framework for supply chain performance measurement. *International Journal of Production Economics* 87(3): 333–347.
35. Li, D. and O'Brien, C. (1999). Integrated decision modelling of supply chain efficiency. *International Journal of Production Economics* 59(1–3): 147–157.
36. Talluri, S. and Baker, R. C. (2002). A multi-phase mathematical programming approach for effective supply chain design. *European Journal of Operational Research* 141(3): 544–558.
37. Claro, D. P., Hagelaar, G., and Omta, O. (2003). The determinants of relational governance and performance: How to manage business relationships? *Industrial Marketing Management* 32(8): 703–716.
38. Gunasekaran, A., Patel, C., and Tirtiroglu, E. (2001). Performance measures and metrics in a supply chain environment. *International Journal of Operations & Production Management* 21(1/2): 71–87.
39. Murphy, G. B., Trailer, J. W., and Hill, R. C. (1996). Measuring performance in entrepreneurship research. *Journal of Business Research* 36: 15–23.
40. Berry, D. and Naim, M. M. (1996). Quantifying the relative improvements of redesign strategies in P.C. supply chain. *International Journal of Production Economics* 46/47: 181–196.
41. Thonemann, U. W. and Bradley, J. R. (2002). The effect of product variety on supply chain performance. *European Journal of Operational Research* 143: 548–569.
42. Agarwal, A., Shankar, R., and Tiwari, M. K. (2006). Modeling the metrics of lean, agile and leagile supply chain: An ANP-based approach. *European Journal of Operational Research* 173: 211–225.
43. Min, H. and Zhou, G. (2002). Supply chain modeling: Past, present and future. *Computers & Industrial Engineering* 43: 231–249.
44. Govindan, K., Diabat, A., and Popiuc, M. N. (2012). Contract analysis: A performance measures and profit evaluation within two-echelon supply chains. *Computers & Industrial Engineering* 63: 58–74.

45. Paksoy, T., Bektaş, T., and Özceylan, E. (2011). Operational and environmental performance measures in a multi-product closed-loop supply chain. *Transportation Research Part E* 47: 532–546.
46. Correia, I., Melo, T., and Saldanha-da-Gama, F. (2013). Comparing classical performance measures for a multi-period, two-echelon supply chain network design problem with sizing decisions. *Computers & Industrial Engineering* 64: 366–380.
47. Li, X., Gu, X. J., and Liu, Z. G. (2009). A strategic performance measurement system for firms across supply and demand chains on the analogy of ecological succession. *Ecological Economics* 68: 2918–2929.
48. Cai, J., Liu, X., Xiao, Z., and Liu, J. (2009). Improving supply chain performance management: A systematic approach to analyzing iterative KPI accomplishment. *Decision Support Systems* 46: 512–521.
49. Kroes, J. R. (2007). Outsourcing of supply chain processes: Evaluating the impact of congruence between outsourcing drivers and competitive priorities on performance. PhD thesis, Georgia Institute of Technology, Atlanta, GA.

13 Simulation in Various Disciplines

13.1 INTRODUCTION

Simulation is a copy of a real-world situation with a target to find problems in the system so as to rectify them. Naturally, such imitations are possible for any kind of system if the proper tools are available. Thus, this chapter presents an overview of the application of simulation in various disciplines.

However, there are so many disciplines in which the concept of simulation can be applied that it would take an entire other book to describe all of them. Some of the significant areas where simulation studies can be applied include various engineering disciplines such as electronic and communication engineering, electrical engineering, mechanical engineering, production engineering, architectural engineering, chemical engineering, civil engineering, mining engineering, environmental engineering, aerospace engineering, ocean engineering, and textile engineering. In addition, there are other areas where simulation studies can be conducted. Such areas include physics, chemistry, biology, biochemistry, biotechnology, nanotechnology, and geography among numerous others. Since it is not possible to discuss in detail about all the disciplines in the whole world of knowledge, this chapter emphasizes some disciplines where a significant number of simulation tools are available. Thus, Section 13.2 discusses simulation in electronics engineering.

13.2 SIMULATION IN ELECTRONICS ENGINEERING

Simulation in the electronics field is basically the simulation of electronic circuits. The various areas of electronics where simulation studies can be conducted can include any of the following:

- Design of analog electronic circuits
- Design of digital electronic circuits
- Semiconductor manufacturing
- Signal analysis
- Signal timings
- Mixed-signal electronic circuit simulation
- Symbolic circuit simulation
- Study of binary values

- Connections among various parts of a circuit
- Communication among circuit components
- Simulation of functions of metal–oxide–semiconductor field-effect transistor (MOSFET) and diode

An electronic circuit can have many components and thus many properties. Almost all these properties can be simulated. There are a vast number of simulators available for simulating electronics applications and circuits. Some of these simulation tools are listed in Sections 13.2.1 and 13.2.2.

13.2.1 OPEN ACCESS SOFTWARE

- Electric very-large-scale integration (VLSI) design systems [for drawing layout of integrated circuits (ICs)]
- GPSim (for analog circuits)
- Simulation programs with IC emphasis (SPICE) (for analog circuits)
- KTechLab (for digital circuits)
- TkGate (for digital circuits)
- DC/AC Lab online circuit simulator (for mixed circuits)
- GNU Circuit Analysis Package (for mixed circuits)
- GeckoCIRCUITS (for mixed circuits)
- Quite Universal Circuit Simulator
- GnuCap (for MOSFETs and diodes)
- CircuitLogix (for switches, digital ICs, linear ICs, FETs, transistors, relays, signal generators, photo diodes, semiconductors, and motors)
- MultiSim (for circuit design and circuit analysis)
- XSpice
- QUCS

13.2.2 PROPRIETARY SOFTWARE

- SimOne (for analog circuits)
- PartSim (for analog circuits)
- LTSpice (for analog circuits)
- SapWin (for analog circuits)
- CPU Sim (for digital circuits)
- Logisim (for digital circuits)
- TINA-TI (for digital circuits)
- Micro-Cap (for mixed circuits)
- Symbulator (for symbolic circuits)
- Syrup (for symbolic circuits)
- PCSPICE
- SiMetrix
- PowerSim (for dynamic system simulation)

13.3   SIMULATION IN CHEMICAL ENGINEERING

The discipline of chemical engineering also has a vast number of tools and techniques as evident from the existing literature. However, the areas where simulation may be applied in this discipline include the following:

- Chemical properties of various materials
- Energy balances
- Chemical environment modeling
- Distillation
- Process flows
- Flowsheet simulation of solid processes
- Design of chemical plant
- Temperature and pressure control of a chemical reaction

A vast number of tools are used for the simulation purposes in this field. Some of the more significant ones are as follows:

- ASCEND
- Dymola
- COMSOL Multiphysics
- EcosimPro
- COCO Simulator
- DWSIM
- MapleSim
- ProMax
- OLGA
- SIMULIS
- SolidSim
- Petro-Sim
- Usim Pac
- SuperPro
- ProSimulator
- OpenModelica
- Omegaland
- Mobatec Modeller
- MiMic
- IDEAS
- ICAS
- INDISS
- IiSE Simulator

The above-mentioned list of simulators includes both open-source simulation packages and proprietary tools.

13.4   SIMULATION IN AEROSPACE ENGINEERING

The application of simulation in aerospace engineering is not as vast as it is in the other disciplines. However, the areas in aerospace applications where simulation studies have been conducted include the following:

- Solid propellant ballistics simulation
- Thermodynamic and transport properties
- Rocket performance
- Thermal balance
- Safety assessment
- Liquid rocket engine performance
- Chemical equilibrium analysis
- Rocket engine simulation
- Vibration analysis
- Turbine engine design
- Functionalities in aerodynamics
- Orbit propagation
- Orbit trajectory problems
- Preliminary structural design
- Workflow simulation
- Aerospace manufacturing
- Vehicle simulation
- Combustion equilibrium analysis
- Fluid flow design
- Heat transfer
- Flight dynamic analysis
- Satellite development and verification
- Aerospace system verification
- Simulation of wave propagation
- Maintenance simulation

There are a number of tools for modeling and simulation in this area. Some of the more significant ones are as follows:

- BurnSim
- REDTOP-Lite
- TURBINE-PAK
- ASTOS
- DG GL-Predict
- Celestia
- FlightSight
- Satellite Tool Kit
- Trajectory Planner
- ANSIS Professional NLT
- ASTROS

- Beam Analyzer
- Composite Analyzer
- MSC Nastran
- CHEMKIN-PRO
- Gaseq
- Sentry
- Thermal Desktop
- Caedium
- Cart3D
- Aerodynamic preliminary analysis system (APAS)
- Digital DATCOM
- SC/Tetra
- OpenFOAM
- STREAM
- XFOIL
- ISight

13.5   SIMULATION IN CIVIL ENGINEERING

Simulation in civil engineering has also seen many applications in various disciplines. Some of the areas in this discipline are as follows:

- Cartography
- Architecture
- Land planning
- Land surveyors
- Analysis of stratified slope stability
- Land developments
- Water projects
- Vehicle swept path analysis
- Vehicle maneuvers for transportation design
- Water distribution planning
- Groundwater modeling
- Terrain modeling
- Surface water modeling
- Wastewater treatment
- Civil construction
- Drainage system design
- Roadway design
- Concrete multistoried building design
- Dynamic analysis of concrete buildings in seismic areas
- Road designs
- Traffic accident studies
- Highway design
- Operational forecasting of urban and rural catchments
- Storm drain inlets

- River management
- Coastal engineering
- Landfall designing
- River channels and flood plains
- Building site design
- Bridge design

There is a very vast number of software packages and tools for such a vast number of areas. Some of the more significant tools among them are as follows:

- ArtiosCAD
- ASCAD
- ASPEN
- AeroTURN
- AUTOCAD Civil 3D
- CivilFEM
- EDISIS 2000
- Civil Designer
- ELPLA
- FloodWorks
- GEOPAK
- GuidSIGN
- HYDRONET
- Ezicad
- IcoMap
- Geocomp Systems
- Inlet Master
- InfoWater
- ParkCAD
- InRoads
- RISA-3D
- RockWare
- SOLAIO 2000
- Surveyor 3D
- WaterGEMS
- Topocad

13.6   SIMULATION IN OTHER DISCIPLINES

A vast number of disciplines and related simulation software packages are available in the existing literature. Mentioning and describing all these software packages is just not practical from the aspect of its vastness. Thus, Table 13.1 provides a glimpse of the various software packages in numerous other disciplines.

TABLE 13.1
Software Packages in Various Disciplines

Software	Developers/Relevant Authors	Year	Function	Application Area
XSophe-Sophe-XeprView® [1]	Hanson, G. R., Gates, K. E., Noble, C. J., Griffin, M., Mitchell, A., and Benson, S.	2004	To calculate Hamiltonian parameters from wave electron spectra from meta ion centers	Inorganic biochemistry
ProcessModel Release 3.0™ [2]	ProcessModel Inc., Provo, Utah	2000	To simulate business processes	Business process
SummerPack [3]	Orsini, J.-P. G.	1990	To assist farmers and agricultural advisors for management of sheep during summer and autumn	Agricultural systems
WinQSB [4]	Amariei, O. I., Frunzaverde, D., Popovici, G., and Hamat, C. O.	2009	To train professionals with various technologies such as PERT/CPM and facility location	Training of industry professionals; behavioral sciences
ATLAS [5]	Marshall, Z., the ATLAS collaboration	2009	To design layout and to implement inert materials for nuclear detectors	Nuclear physics
SIMION	Don C. McGilvery	1970	To calculate electron fields and trajectories of charged particles	Optics and electronics
AMESim	Imagine S.A., a company that was acquired in June 2007 by LMS International, Leuven, Belgium	1987	To analyze and predict mechatronics systems	Systems engineering
PKUDDS (Peking University Drug Design System) [6]	Hou, T. and Xu, X.	2001	To design molecular structure for drugs	Molecular design and modeling
ALLAN [7]	Boulkroune, K., Candau, Y., Piar, G., and Jeandel, A.	1995	To design and simulate dynamic system	Dynamic systems, especially the thermal systems
BetaSim [8]	Kreutzer, W. and Østerbye, K.	1998	To develop layered design for discrete event simulation software by using beta programming language	Queuing and other networks

(Continued)

TABLE 13.1

(Continued) Software in Various Disciplines

Software	Developers/Relevant Authors	Year	Function	Application Area
DYSMPS2 [9]	Dangerfield, B.	1992	To make strategic policy and for group decision support system	System dynamics
MyM [10]	Beusen, A. H. W., de Vink, P. J. F., and Petersen, A. C.	2011	To develop integrated design of mathematical models, execution, data analysis, and visualization to facilitate communication among model builders, analysts, and decision makers	Environmental sectors
SAMT (Spatial Analysis and Modeling Tool) [11]	Wieland, R., Voss, M., Holtmann, X., Mirschel, W., and Ajibefun, I.	2006	To integrate models from different branches of study, such as economics and ecology	Geographic landscapes
SPROM [12]	Cai, C., Lin, M., Chen, Z., Chen, X., Cai, S., and Zhong, J.	2008	To simulate nuclear magnetic resonance spectra and images	Quantum physics
FRASTA (Fracture Surface Topography Analysis) [13]	Cao, Y. G. and Tanaka, K.	2006	To simulate the process of formation of crack and to calculate the relevant parameters such as crack opening angle and crack-tip opening angle	Topography analysis
FLUENT	ANSYS, Canonsburg, Pennsylvania	1996	To simulate turbulence, heat transfer, and reactions among various fluids, such as air flow over aircraft wing, bubble columns, and blood flow	Fluid dynamics, semiconductor manufacturing, and wastewater treatment plants
FDS (Fire Dynamics Simulator)	The National Institute of Standards and Technology (NIST), US Department of Commerce, and VIT Technical Research Center of Finland	2000	To simulate low-speed flows, such as smoke or heat from fires	Evolution, study, and investigation of fire; fire protection technology
TRNSYS [14]	Beckman, W. A., Broman, L., Fiksel, A., Klein, S. A., Lindberg, E., Schuler, M., and Thornton, J.	1994	To simulate and model solar energy system	Renewable energy system

(Continued)

TABLE 13.1
(Continued) Software in Various Disciplines

Software	Developers/Relevant Authors	Year	Function	Application Area
Genesis [15]	Liu, Y., Szymanski, B. K., and Saifee, A.	2006	To simulate parallel network to increase scalability and efficiency; the main emphasis is on the complexity and dynamics of a network	Computer networks, especially the Internet
Easy5 [16]	Fritz, N., Elsawy, A., Modler, K.-H., and Goldhahn, H.	1999	To simulate hydraulics, various control systems, refrigeration, lubrication, or fuel systems	System engineering for aircraft, vehicles, agricultural equipment, and other complex systems
WITNESS [17]	Clark, M. F.	1991	To simulate and analyze various types of manufacturing systems; the models objects can be defined, designed, and changed interactively	Manufacturing systems
BALANCE [18]	Wesseling, J. G.	2006	To simulate the flow of soil moisture and to simulate the interaction among the various components for water balance	Environmental study

Source: AMESim, http://www.lmsintl.com/LMS-Imagine-Lab-AMESim; ANSYS, http://www.ansys.com/Products/Simulation+Technology/Fluid+Dynamics/Fluid+Dynamics+Products/ANSYS+Fluent; FDS, http://code.google.com/p/fds-smv/.

13.7 SOME SELECTED SIMULATION PACKAGES

This section provides brief descriptions of a few simulation software packages in various disciplines. The features listed under each software package represent the description of the package itself.

13.7.1 FLUENT

- It quickly prepares product/process geometry for flow analysis without tedious rework.
- It avoids duplication through a common data model that is persistently shared across physics beyond basic fluid flow.

- It easily defines a series of parametric variations in geometry, mesh, physics, and postprocessing, enabling new automatic computational fluid dynamics (CFD) results for that series with a single mouse click.
- It improves product/process quality by increasing the understanding of variability and design sensitivity.
- It easily sets up and performs multiphysics simulations.
- It can be applied to solve flow problems using unstructured meshes.
- It can handle all types of flow models such as steady-state or transient flows, incompressible or compressible flows, laminar or turbulent flows, Newtonian or non-Newtonian flows, and ideal or real gases.
- It can be applied to industrial equipment where optimizing the heat transfer is difficult.
- It is very appropriate for multiphase modeling technology.
- It can provide a good framework for combustions and chemical reactions for flow of fluids.
- It is capable of computing noise from unsteady pressure fluctuations.
- The effect of solid motion on fluid flow can be modeled.
- It has high-performance computing capabilities.
- It supports parallel processing.
- It is capable of optimizing shapes, and thus it can adjust geometric parameters.
- It has a variety of material modeling options, making a variety of solutions possible.
- It is easy to customize depending on the needs of users.
- It has effective graphics, animation, and report-making capabilities that can be used in preparing postprocessing reports.

13.7.2 TRNSYS

- Also known as transient system simulation, it is an energy simulation software package.
- It has the ability to interface with other similar simulation software packages.
- It is a component-based software package.
- It is used to simulate the behavior of transient systems.
- It is also flexible enough to simulate traffic flow and biological processes.
- It is composed of two main parts— an engine and a library of components. The engine is also called the kernel.
- The function of the kernel or engine is to process an input file, solve the problem, determine convergence, and plot the system variables. The kernel can also determine the thermophysical properties, perform linear regression analysis, and interpolate external data files.
- The library that contains 150 different models. These models consist of library routines for pumps, multizone buildings, wind turbines, eletrolyzers, weather data processors, and economics routines.
- The models in TRNSYS are so flexible that users can modify the existing components and build their own components based on their needs.

13.7.3 EASY5

- It is used for virtual product development.
- It is a graphics-based tool that can be used to simulate and design dynamic systems using various difference equations, differential equations, algebraic equations, and nonlinear simulation tools.
- It can handle complex engineering tasks involving complex dynamic systems.
- It is especially applicable to aerospace and defense equipment, automotive equipment, off-highway equipment, and other heavy machinery.
- It is an open-source software that can be connected to a wide range of software and hardware tools used in computer-aided engineering.
- It has several function blocks, such as summers, dividers, lead–lag filters, and integrators, and special system components, such as valves, actuators, heat exchangers, gears, clutches, engines, pneumatics, and flight dynamics.
- Its various tools used for model building include nonlinear simulation, steady-state analysis, linear analysis, control system analysis, data analysis, and plotting of data.
- It supports automatic generation of source code.
- It provides a user-friendly user interface.
- Each component in this model consists of parametric data to define the system properties and equations to define the dynamics of the system.

13.7.4 GENESIS

- Also known as general neural simulation system, it is a simulation software that simulates neurobiological systems.
- It basically models the anatomical and physiological characteristics of neural systems.
- It quantifies the physical framework of the nervous system in such a way that it facilitates an easy understanding of the physical structure of the nervous system.
- It facilitates parallel modeling of networks on multiple-instruction, multiple-data computers.
- It consists of three modules: script files, a graphical user interface, and a command shell.
- It consists of an interpreter that interprets script language constructs. There is a set of library routines.
- Its application lies in biological systems—it usually simulates brain structures. It is also used in biomedical applications.

13.7.5 BetaSim

- It provides a layered design for a discrete event simulation framework.
- It is based on beta simulation language.

- It is a powerful tool to model classical patterns of interaction.
- It is particularly useful for modeling queuing systems.
- It consists of several layers such as a scenario layer, a synchronization layer, a sampling layer, a monitor layer, an instrumentation layer, and a beta layer.

13.7.6   AMESim

- It is a simulator suitable for multidomain systems.
- It is especially applicable for measuring the performance of mechatronics systems.
- Nonlinear time-dependent analytic equations are used to develop models with AMESim.
- It can model various system behaviors such as hydraulic, pneumatic, thermal, electrical, and mechanical behaviors.
- Modeling in AMESim is done in four steps: sketch mode, submodel mode, parameter mode, and run mode.
  - In the sketch mode, the different components of the model are linked.
  - In the submodel mode, the submodel associated with each component is chosen.
  - In the parameter mode, the parameters for each submodel are decided.
  - In the run mode, the simulation model is run or executed and the results obtained are analyzed.
- It is compiled between the submodel mode and the parameter mode.

13.7.7   SimOne

- It was developed by SIMONE Research Group, Inc., Czech Republic, and LIWACOM Informationstechnik GmbH, Essen, Germany.
- It is a simulation software for the optimization of gas transport and distribution.
- It is applied for pipeline system design, gas dispatching, marketing analysis, and operation planning.
- It is capable of handling complex pipeline networks.
- It considers all the relevant tools, equipment, and components for pipeline networks such as pipes, compressor stations, pressure reducers, valves, blending stations, heaters/coolers, and storage plants.
- It is capable of performing three types of simulations—simulation of steady states and dynamic processes, optimization of gas transport and distribution, and simulation of network control and dispatching.
- It provides an object-oriented interface through which all the above three components interact among one another.
- Its graphical user interface provides a powerful visualization of the simulation model.

13.7.8   LogiSim

- It is a circuit simulation software.
- It uses a graphical user interface to visualize the simulated circuit designs.

- It is mostly used by and for students to show the designing and experimentation with digital circuits in simulation.
- It is flexible enough to let the users modify the circuit designs during simulation.

13.7.9 DWSIM

- It is an open-source chemical process simulator.
- It is written in Visual Basic.NET software.
- It can be used to simulate thermodynamic systems, reacting systems, and petroleum characterization systems.
- It supports multivariate optimization and sensitivity analysis.
- It provides a process flowsheet diagram (PFD) drawing interface.
- It supports chemical reactions and reactors.
- This model consists of a single executable for both .NET and mono runtimes.
- It provides an Excel interface for thermodynamic calculations.
- It consists of rigorous distillation and absorption column models.
- It includes IronPython/IronRuby scripting and is extensible through IronPython and Lua scripts and plugins.

13.8 CONCLUSION

In this chapter, simulation studies in some other disciplines have been provided. Although there are numerous disciplines where some kind of simulation study is applied, discussion of simulation studies in all the areas is not possible in a single chapter. Therefore, this chapter has discussed the application areas of simulation studies and the respective simulation tools and software packages in electronics, chemical engineering, aerospace engineering, and civil engineering. Each of these fields uses a vast number of simulation tools and software packages that are easy to obtain and learn, since many are open-source simulation tools.

REFERENCES

1. Hanson, G. R., Gates, K. E., Noble, C. J., Griffin, M., Mitchell, A., and Benson, S. (2004). XSophe-Sophe-XeprView®. A computer simulation software suite (v. 1.1.3) for the analysis of continuous wave EPR spectra. *Journal of Inorganic Biochemistry* 98(5): 903–916.
2. Milton, S. M. (2000). ProcessModel Release 3.0™: Business process simulation software. *Long Range Planning* 33(3): 440–442.
3. Orsini, J.-P. G. (1990). "SummerPack," a user-friendly simulation software for the management of sheep grazing dry pastures or stubbles. *Agricultural Systems* 33(4): 361–376.
4. Amariei, O. I., Frunzaverde, D., Popovici, G., and Hamat, C. O. (2009). WinQSB Simulation Software—A tool for professional development. *Procedia—Social and Behavioral Sciences* 1(1): 2786–2790.
5. Marshall, Z. (2009). The ATLAS simulation software. *Nuclear Physics B—Proceedings Supplements* 197(1): 254–258.

6. Hou, T. and Xu, X. (2001). A molecular simulation software package—Peking University Drug Design System (PKUDDS) for structure-based drug design. *Journal of Molecular Graphics and Modelling* 19(5): 455–465.

7. Boulkroune, K., Candau, Y., Piar, G., and Jeandel, A. (1995). Validation of a building thermal model by using ALLAN. Simulation software. *Energy and Buildings* 22(1): 45–57.

8. Kreutzer, W. and Østerbye, K. (1998). BetaSim: A framework for discrete event modelling and simulation. *Simulation Practice and Theory* 6(6): 573–599.

9. Dangerfield, B. (1992). The system dynamics modeling process and DYSMPA2. *European Journal of Operational Research* 59(1): 203–209.

10. Beusen, A. H. W., de Vink, P. J. F., and Petersen, A. C. (2011). The dynamic simulation and visualization software MyM. *Environmental Modelling & Software* 26(2): 238–240.

11. Wieland, R., Voss, M., Holtmann, X., Mirschel, W., and Ajibefun, I. (2006). Spatial Analysis and Modeling Tool (SAMT): 1. Structure and Possibilities. *Ecological Informatics* 1(1): 67–76.

12. Cai, C., Lin, M., Chen, Z., Chen, X., Cai, S., and Zhong, J. (2008). SPROM—An efficient program for NMR/MRI simulations of inter- and intra-molecular multiple quantum coherences. *Comptes Rendus Physique* 9(1): 119–126.

13. Cao, Y. G. and Tanaka, K. (2006). Development of FRASTA simulation software. *Acta Metallurgica Sinica (English Letters)* 19(3): 165–170.

14. Beckman, W. A., Broman, L., Fiksel, A., Klein, S. A., Lindberg, E., Schuler, M., and Thornton, J. (1994). TRNSYS: The most complete solar energy system modeling and simulation software. *Renewable Energy* 5(1–4): 486–488.

15. Liu, Y., Szymanski, B. K., and Saifee, A. (2006). Genesis: A scalable distributed system for large-scale parallel network simulation. *Computer Networks* 52(12): 2028–2053.

16. Fritz, N., Elsawy, A., Modler, K.-H., and Goldhahn, H. (1999). Simulation of mechanical drives with Easy5. *Computers & Industrial Engineering* 37(1/2): 231–234.

17. Clark, M. F. (1991). WITNESS: Unlocking the power of visual interactive simulation. *European Journal of Operational Research* 54(3): 293–298.

18. Wesseling, J. G. (1993). BALANCE: A package to show the components of the water balance of a one-dimensional soil profile in time. *Environmental Software* 8(4): 247–253.

14 Simulation of Complex Systems

14.1 INTRODUCTION

The study of complexity is a fascinating area that needs a multilevel and multidomain approach that gives microlevel details of the system under study. Microlevel simulation is an increasingly interesting and challenging area of research for researchers and practitioners all over the world. In general, problems that are difficult to understand and solve are known as complex problems. But this definition is an extremely vague definition because the level of understanding and solution capability depend on individual capability.

According to Sterman [1], there are two types of complexities: detail complexity and dynamic complexity. He defines these types of complexity as follows:

> Most people think of complexity in terms of the number of components in a system or the number of combinations one must consider in making a decision. The problem of optimally scheduling an airline's flights and crews is highly complex, but the complexity lies in finding the best solution out of an astronomical number of possibilities. Such needle-in-a-haystack problems have high levels of combinatorial complexity (also known as detail complexity). Dynamic complexity, in contrast, can arise even in simple systems with low combinatorial complexity. [...] Dynamic complexity arises from the interactions among the agents over time.

However, in general, in cases with complex details, there are too many interacting components in the system. Thus, the overall outcome of this type of systems depends on the extent of collaboration among these components.

According to Day [2], "a dynamical system is complex if it endogenously does not tend asymptotically to a fixed point, a limit cycle or an explosion." This definition indicates the existence of a strange attractor. According to Chaos Theory [3], these strange attractors force a system to lie within certain ranges.

The science of complexity takes into account both the chaotic behavior and the well-defined characteristics of the system. There is also the concept of a complex adaptive system (CAS) that emphasizes adaptation and thus evolution to the system with changed behavior. A CAS is a nonlinear, complex, and highly interactive system. However, before knowing the details of analyzing and controlling such complex systems, we must know the advantages and disadvantages of simple systems so as to understand the importance and reality of complex systems. Thus, Section 14.2 discusses the advantages and disadvantages of simple systems.

14.2   ADVANTAGES AND DISADVANTAGES OF SIMPLE SYSTEMS

Without a clear understanding of the advantages and disadvantages of simple systems, it is difficult to understand the importance of a complex system. The characteristics and advantages of a simple system can be delineated through the following points:

1. It is easier to understand.
2. It is easier to implement and analyze.
3. It is easier to modify because of the fewer number of components and thus fewer interactions compared to a complex system.
4. It is easier to reduce the running time of a simple simulation study because the components of the system are easier to analyze and modify.

The above advantages automatically indicate the disadvantages as well. Thus, the disadvantages of a simple system are as follows:

1. Most simple models do not represent the reality. This means that most real-life problems are complex in nature. Thus, simple problems may be unrealistic at times.
2. An oversimplified model may lose validity. Thus, complexity should be incorporated in a real-life problem to the maximum accepted level, that is, a system may be complicated if required.
3. Most simple problems are based on a number of assumptions. This means that, in most cases, to solve problems of high complexity, we reduce the problem to a simpler one that deflects the problem far from reality. For example, in many cases in the existing literature, it is seen that to solve a nonlinear system, we reduce it to a linear system that is much easier to solve than a complex nonlinear problem. This abstraction happens after a lot of assumptions that make the reduced problem unrealistic in most cases.
4. Simpler models are often obtained by reducing the scope, which reduces the extent of generalization for the solution of the problem. For example, while solving a supply chain problem, in many cases, we assume serial supply chains of fewer echelons instead of a more realistic supply chain network. This simplification reduces the scope of solution of a proposed problem.

The problem of complexity may be faced primarily by taking the following simpler approaches in many cases:

1. Instead of starting with a complex problem at the very beginning, a simpler problem can be taken at the start. Complexity can then be added step by step. In this way, the solution methodology of the simple problem sometimes becomes helpful to the solution of a problem with added complexity.
2. The use of hierarchy is often thought of as a way to simplify or deal with a complex model. Although such an approach may not be applicable to all types of problems, this particular approach may be effective for some problems.

However, in most cases, oversimplification results in an unrealistic problem. In those cases, a different approach is required. Thus, Section 14.3 presents approaches to deal with such complex problems.

14.3 EFFECTIVE TOOLS TO SIMULATE AND ANALYZE COMPLEX SYSTEMS

A complex system faces two main challenges among others: nonlinearity of the complex system and a lot of details to be presented. Some of the significant difficulties in solving a complex problem are as follows:

1. It is generally nonlinear in nature.
2. It has a significant number of interactive components.
3. It is difficult to handle with analytical systems.
4. It is difficult to decompose into subsystems.
5. It is difficult to reduce by general laws applicable to most systems.
6. Models based on nonlinear differential equations are very difficult to solve.

However, according to the level of abstraction, a model can be classified into an abstract model, a prototype model, or a simulation model. Simulation of a complex problem is a better way to find a solution. But it has a big disadvantage, which lies in the fact that a simulation model is often dependent on the context of the problem under study. Thus, finding a generalized solution of a complex problem by simulation is difficult. However, on the positive side, a simulation study can deal with complex problems. Thus, the contextual problem can be solved by running a simulation model for various contexts.

Simulation of a complex problem may be done by applying several methodologies. Application of a particular methodology depends on the type of problem under study. However, the following methods have been found to be effective in handling complex systems.

1. *Agent-based modeling and simulation.* Complex systems are often characterized by a significant number of interacting components. Such interacting components may be represented by "agents."

   A concise definition of agents is difficult to find. In sum, it can be said that agents are self-contained, autonomous, adaptive, interacting components of a system, and they are capable of learning from their environment and making independent decisions. A true agent with perfectly independent decision making is yet to be implemented, and the implementation is also dependent on the existing technology. Agents are generally defined as software components, and there lies the devil.

   A software component is always programmed, and thus the agents are also programmed. Thus, all the behaviors and characteristics of the agents are programmed. Therefore, the independent decision making is also programmed because the context of the decision making is predicted and decided beforehand.

The "self-contained" behavior of an agent means that all the activities and properties are defined into a compact single agent. The behaviors are defined by functions, and various properties are defined by various programming constructs. Thus, all the behavioral functions are defined somewhere in the agent system. The behavior may be contained in a single agent or can be shared and thus may be a common behavior among some agents.

The "autonomous" property indicates that the agents are independent bodies capable of moving around if required and making individual decision. But the movement of an agent depends on the stimulating situations that are always defined, and the scope of movement is also defined in an agent system.

The "adaptive" behavior means that the agents can adjust to the changes in a dynamic environment. Such adjustments indicate the modification of its behavior, which is also defined beforehand, and the possible changes in a system are also predicted while developing an agent system.

From the above discussion, it is clear that a perfect agent system is difficult to build. However, agent-based systems are certainly the application of artificial intelligence since we endeavor to apply the natural intelligence of living beings into the agents using some kind of programming method.

However, depending on the application areas, there are several types of agents, such as biological agents, robotic agents, or computational agents. Based on the method of working, the agents can be classified into collaborative agents, interface agents, mobile agents, information agents, reactive agents, hybrid agents, and heterogeneous agents.

The main advantages of applying agent-based implementations of complex systems are that the different types of components in complex systems may each be represented by separate agents and the interaction among them can also be programmed. However, programming such systems demands proficiency and skills on the part of both the designers and the programmers. The implementation also depends on the available technology. Chapter 9 discusses agent-based systems in detail.

2. *Artificial neural network (ANN).* The use of an ANN is an effective way to model adaptive complex systems. ANNs are based on human brain activities. This branch of study basically mimics the functionalities of neurons, the brain cells. They can be adapted to input patterns.

The implementation of an ANN is dependent on the quantity of data. Its effectiveness of ANN depends on its learning from the available data. Thus, it will be effective to those applications where significant amounts of data are available on hand. The entire set of data is divided into two different sets: One of these sets is used for learning, that is, to train the network, and the other set is used for getting the required result.

There are two basic kinds of learning in neural networks—supervised learning and unsupervised learning. Unsupervised learning does not teach the agents to react properly to input patterns without considering feedback, whereas supervised learning can teach agents even without considering feedback.

ANNs are a vast field of study. There are a number of aspects of ANNs to be considered while modeling a system. ANNs include the basic concepts of single-input and multi-input neurons and transfer functions. For complex systems, multi-input neurons are appropriate. Among the network architectures, there are multilayered neuron architecture and recurrent network architecture. Among the various types of networks, the significant ones are as follows:

a. Perceptron network
b. Hamming network
c. Hopfield network
d. Adeline network
e. Competitive network
f. Rossberg network

In addition, there are several learning rules in cases of ANNs. Some of the learning rules are as follows:

a. Perceptron learning
b. Supervised Hebbian learning
c. Widrow–Hoff learning
d. Associative learning
e. Learning vector quantization (LVQ) learning

The performance optimization of a neural network can be done by the steepest descent method, the Newton's method, and the conjugate gradient method. However, any of these techniques can be applied to implement and analyze a complex system, depending on the type of system under consideration. For a detailed understanding of ANNs, the work by Hagan et al. [4] and Rojas [5] may be consulted.

3. *Genetic algorithm (GA)*. It is a type of metaheuristic optimization algorithm. It is an inexact optimization algorithm. This means that it cannot guarantee an optimal solution to a problem, and at the same time, there is no way to invalidate its solution's optimality. It is a heuristics algorithm and therefore a trial-and-error approach to solve a problem. GAs can be used in solving complex problems. Thus, the existing literature shows a large number of applications of GAs in solving complex problems.

GAs have been developed based on the natural concept of genetics. GAs are a population-based algorithm, which means that a population of solutions is obtained as a result of a GA in most cases. A population is made up of a number of individuals where each individual is actually a chromosome.

Each chromosome represents a solution for a GA. Each solution or chromosome is evaluated through several iterations in a GA. Each chromosome is composed of several genes. The value of a gene at a particular position of a chromosome is called an "allele." Allele values can be either binary or real.

The GA is implemented through the use of several genetic operators. There are mainly three operators in a GA—crossover operator, mutation operator, and selection operator. There are numerous algorithms for each of these operators. Examples of crossover operators include one-point crossover, *m*-point crossover, uniform crossover, partially matched crossover, and simulated binary crossover. There are also numerous mutation algorithms as evident from the existing literature.

In a GA, at first, a population of chromosomes is generated randomly. Then, a set of fit chromosomes is selected from the population. The selected set contains the best chromosomes from the population. The crossover and mutation operators are used on the chromosomes of this selected set, which generates a number of offspring. After this, the offspring population is combined with the original population to form an intermediate population. Next, the best chromosomes from this intermediate population are selected to fill the original population size. The population size is generally assumed to be fixed. The above process repeats up to a maximum number of generations. For a detailed understanding of GAs, the work by Reeves and Rowe [6] may be consulted.

4. *Other nature-based algorithms.* There are also other nature-based algorithms that have the capability of handling complex problems. All these algorithms are inexact optimization algorithms like GAs. Some of such significant algorithms are as follows:
   a. Particle swarm optimization simulated annealing
   b. Differential evolution algorithm
   c. Ant colony optimization algorithm
   d. Artificial immune system
   e. Tabu search
   f. Honey-bee mating algorithm
   g. High-dimensional objective genetic algorithm
   h. Gene expression algorithm
   i. Frog leaping algorithm
   j. Symbiotic evolutionary algorithm
   k. Membrane algorithm
   l. Bacteria foraging algorithm
   m. Cultural algorithm
   n. Firefly algorithm
   o. Cuckoo search algorithm
   p. Gravitational search algorithm

   For a detailed understanding of these algorithms, the work by Bandyopadhyay and Bhattacharya [7] may be consulted.

5. *System dynamics.* It is a behavioral simulation approach. Unlike the previous algorithms, system dynamics is more a tool for simulation and analysis of behavioral systems. A detailed simulation approach with system dynamics is provided in Chapter 20.

Some of the other tools for simulating complex systems include cause-and-effect diagrams and cellular automata. Simulations with cellular automata are discussed in Chapter 15 in detail.

14.4 CONCLUSION

This chapter has given a brief overview of simulation of complex systems. Particular emphasis has been given on the modeling and simulation of complex systems, rather than going into the mathematical details of complex systems. A number of simulation tools such as an agent-based approach, an ANN, nature-based algorithms, and system dynamics have been mentioned in this chapter. We hope that this chapter will be an effective reference to the study on the simulation of complex systems.

REFERENCES

1. Sterman, J. D. (2000). *Business Dynamics: Systems Thinking and Modeling for a Complex World*. Singapore: McGraw-Hill.
2. Day, R. H. (1994). *Complex Economic Dynamics, Volume I: An Introduction to Dynamical Systems and Market Mechanisms*. Cambridge, MA: MIT Press.
3. Williams, G. P. (1997). *Chaos Theory Tamed*. Washington, DC: Joseph Henry Press.
4. Hagan, M. T., Demuth, H. B., and Beale, M. (1996). *Neural Network Design*. Beijing, People's Republic of China: China Machine Press.
5. Rojas, R. (1996). *Neural Networks: A Systematic Introduction*. Berlin, Germany: Springer-Verlag.
6. Reeves, C. R. and Rowe, J. E. (2003). *Genetic Algorithms: Principles and Perspectives*. New York: Kluwer Academic Publishers.
7. Bandyopadhyay, S. and Bhattacharya, R. (2013). On some aspects of nature-based algorithms to solve multi-objective problems. In X. S. Yange (Ed.), *Artificial Intelligence, Evolutionary Computing and Metaheuristics: In the Footsteps of Alan Turing*. Berlin, Germany: Springer-Verlag, pp. 477–524.

14.2 CONCLUSION

15 Simulation with Cellular Automata

15.1 INTRODUCTION

In Chapter 14, one of the analysis tools mentioned was cellular automata. Thus, in this chapter, an introduction to simulation with cellular automata is provided. This chapter will not go into the complex symbolic details of automata, but will endeavor to describe it in as friendly a way as possible. The word "automata" is the plural of automaton. It means something that works automatically. To implement an automaton, what we need is a generalized language that may represent the basic logic of operations. There are two types of languages—formal languages and informal languages. To say something in a language, we need alphabets, strings made of alphabets, and words for a particular language.

Thus, different languages have different set of alphabets. In automata, language can be described in several ways: descriptive definitions, recursive definitions, regular expressions, and finite automaton language constructs. Descriptive definitions define the conditions imposed on words. Recursive definitions use the concept of recursion to define a language. For example, if the letter "a" is repeated one or more times, then we can use the + symbol thus: ({a}+). Each of the regular expressions and the finite automaton has its own constructs to define the respective language.

The branch of finite states and automata is a vast area and has its root in the field of theoretical computer science. In this chapter, the emphasis is on cellular automata. Thus, Section 15.2 provides an overview of cellular automata.

15.2 CELLULAR AUTOMATA

Cellular automata are a type of model that shows how elements of a system interact among themselves. Technically, a cellular automaton is usually defined by a lattice L of cells, a finite set of states Q for the cells, a finite neighborhood N, and a local transition function $\delta: Q^N \rightarrow Q$. Generally, L is a regular automaton and each element is called a cell. A cell can be a two-dimensional (2D) square or a three-dimensional (3D) block or any other shape. Consider the 2D cells in Figure 15.1.

In the figure, each of the squares in the matrix is a cell. Suppose a cell can have values of either 0 or 1. The value 1 is represented by black cells and the value 0 is represented by white cells. So, in this figure, each cell has eight neighbors. Thus, each cell's neighborhood is defined. In classical cellular automata, the finite neighborhood N is the same for all cells all of the time. The cell under consideration may not

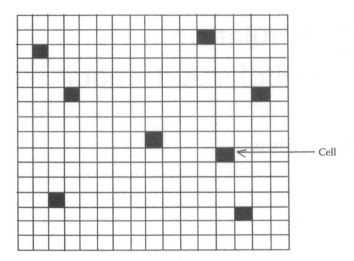

FIGURE 15.1   Cells in cellular automata.

be included in the neighborhood. The characteristics of a cellular automaton can be delineated through the following points:

1. It can be one-dimensional (1D), 2D, 3D, and so on.
2. It may consist of infinitely large grids.
3. The grid in a cellular automaton may grow if required.
4. If cells are distributed, they need to communicate, once per cycle.
5. It is also characterized by synchronous computation.

If the number of cells L is infinite, it becomes very difficult to implement cellular automata in the computer, and if restriction is applied to one specific cellular automaton, this restriction raises the problem of regularity.

Cellular automata can be classified into the following types:

- Partitioned cellular automata: Here, a cell may not use the entire set of states.
- Nondeterministic cellular automata: These are the generalization of standard cellular automata with the additional feature that the local transition rule is a function of δ.
- Asynchronous cellular automata: These are deterministic cellular automata where the definition of global transition has been modified.
- Probabilistic cellular automata: Here, a cell enters a new state $q \in Q$ with a particular probability.
- Heterogeneous cellular automata: The lattice of cells or the neighborhood of cells or the local transitions are not homogeneous.

After introducing the idea of cellular automata, Section 15.3 provides a glimpse of simulation with cellular automata.

15.3 TYPES OF SIMULATION WITH CELLULAR AUTOMATA

Simulation with cellular automata has two aspects, which are as follows:

1. Simulation with cellular automata
2. Simulation of other systems with cellular automata

A cellular automaton C is simulated by another cellular automaton C' if the following conditions are satisfied [1]:

- If both C and C' compute functions f and f', respectively, so that for certain configurations, they generate certain output configurations.
- There is an easily computable function of which, for each configuration, C gives a corresponding configuration $g(C)$ of C'.
- For the input configuration of C, it holds that $g[f(c)] = f'[g(c)]$.

There are numerous algorithms that are used for simulating cellular automata, as is evident from the existing literature. One such algorithm [1] is shown in Figure 15.2.

Cellular automata have also been used for simulations in various disciplines. Examples of such fields are given in Section 15.4.

15.4 APPLICATIONS OF CELLULAR AUTOMATA

A variety of application areas can be identified from the existing literature where cellular automata have been used for simulation purposes. A list of such applications is provided below. Detailed analyses of these applications are not provided because of their complexity.

- Simulation of urban growth processes [2]
- Simulation for the fabrication of integrated circuits [3]
- Simulation of road networks [4]
- Simulation of corrosion mechanisms [5]
- Simulation of solidification morphologies [6]
- Simulation of traffic flows [7]

```
Input: state array c1 [X+1][Y+1] and state array c2 [X][Y];
Algorithm

Initialize Boundaries of c1
For x ← 1 to X do
        For y ← 1 to Y do
                Determine state of c2, c2[x][y] ← New State of neighbors (c1[x –
        1][y – 1],
                                c1[x – 1][y], c1[x – 1][y C 1], c1[x][y – 1], c1[x][y],
                                c1[x][y C 1], c1[x C 1][y – 1], c1[x C 1][y], c1[x C
        1][y C 1])
                EndFor
EndFor
```

FIGURE 15.2 A simulation algorithm for cellular automata.

- Simulation of mammary ductal carcinomas [8]
- Simulation of applications in fluid dynamics [9]
- Simulation of train operations [10]
- Simulation of land use [11]
- Simulation of load balancing problems [12]
- Simulation for improving other algorithms such as the particle swarm optimization algorithm [13]
- Simulation for processing rubber compounds [14]
- Simulation studies in metallurgy [15]

15.5   SOFTWARE FOR CELLULAR AUTOMATA

There are a few software applications available for simulation with cellular automata, as is evident from the existing literature. These software applications are basically languages for cellular automata. The basic components of cellular automata language are (1) lattice, which is generally 2D, and (2) neighborhood of cells.

Many of these languages perform direct compilations of the codes developed. In many other cases, it is a two-stage process. In the first stage, cellular automata programs are compiled into a source code of a different programming language, such as C, and in the second stage, the converted code is again converted to a machine language code.

Some of the famous existing software languages are given as follows:

- JCASim
- CAMEL/CARPET
- CASIM
- CDM/SLANG
- CELLULAR/CELLANG
- HICAL
- CAM SIMULATOR
- CAT/CARP
- CELLAS/FUNDEF
- CEPROL
- LCAU
- SCARLET/SDL
- CAPOW
- CDL
- CELLSIM
- DDLAB
- PECANS/CANL
- SICELA

15.6   CONCLUSION

This chapter has presented a brief overview of simulation with cellular automata. A brief introduction to the theory of automata followed by a brief introduction to the concept of cellular automata has been provided at first. Then, various aspects of simulation with cellular automata along with an algorithm have been presented. A number of application areas followed by a list of languages used for computerized simulations with cellular automata have been shown.

However, the application of cellular automata is yet to be explored in various areas. Cellular automata have a huge prospect to be applied successfully in several branches of study.

REFERENCES

1. Worsch, T. (1999). Simulation of cellular automata. *Future Generation Computer Systems* 16(2/3): 157–170.
2. Santé, I., García, A. M., Miranda, D., and Crecente, R. (2010). Cellular automata models for the simulation of real-world urban processes: A review and analysis. *Landscape and Urban Planning* 96(2): 108–122.
3. Sirakoulis, G. C., Karafyllidis, I., and Thanailakis, A. (2002). A cellular automation methodology for the simulation of integrated circuit fabrication processes. *Future Generation Computer Systems* 16: 639–657.
4. Sun, T. and Wang, J. (2007). A traffic cellular automata model based on road network grids and its spatial and temporal resolution's influences on simulation. *Simulation Modelling Practice and Theory* 15(7): 864–878.
5. Córdoba-Torres, P., Nogueira, R. P., de Miranda, L., Brenig, L., Wallenborn, J., and Fairén, V. (2001). Cellular automation simulation of a simple corrosion mechanism: Mesoscopic heterogeneity versus macroscopic homogeneity. *Electrochimica* 46(19): 2975–2989.
6. De-chang, T. and Weng-sing, H. (2010). Numerical simulation of solidification morphologies of Cu-0.6Cr casting alloy using modified cellular automation model. *Transactions of Nonferrous Metals Society of China* 20(6): 1072–1077.
7. Yi, Z., Houli, D., and Yi, Z. (2007). Modeling mixed traffic flow at crosswalks in micro-simulations using cellular automata. *Tsinghua Science and Technology* 12(2): 214–222.
8. Bankhead III, A., Magnuson, N. S., and Heckendorn, R. B. (2007). Cellular automation simulation examining progenitor hierarchy structure effects on mammary ductal carcinoma *in situ*. *Journal of Theoretical Biology* 246(3): 491–498.
9. Lim, H. A. (1990). Lattice-gas automation simulations of simple fluid dynamical problems. *Mathematical and Computer Modelling* 14: 720–727.
10. Hongsheng, S., Jing, C., and Shiming, H. (2012). Simulation system engineering for train operation based on cellular automation. *Information Engineering and Complexity Science—Part I* 3: 13–21.
11. Wang, S. Q., Zheng, X. Q., and Zhang X. B. (2012). Accuracy assessments of land use change simulation based on Markov-cellular automation model. *Procedia Environmental Sciences* 13: 1238–1245.
12. Hofestädt, R., Huang, X., and Beerbohm, D. (1996). Simulation of load balancing with cellular automata. *Simulation Practice and Theory* 4(2/3): 81–96.
13. Fengxia, Y. and Gang, L. (2012). The simulation and improvement of particle swarm optimization based on cellular automata. *Procedia Engineering* 29: 1113–1118.
14. Bandini, S. and Magagnini, M. (2001). Parallel processing simulation of dynamic properties of filled rubber compounds based on cellular automata. *Parallel Computing* 27(5): 643–661.
15. Lan, Y. Z., Xiao, N. M., Li, D. Z., and Li, Y. Y. (2005). Mesoscale simulation of deformed austenite decomposition into ferrite by coupling a cellular automation method with a crystal plasticity finite element model. *Acta Materialia* 53(4): 991–1003.

16 Agent-Based Simulation

16.1 BACKGROUND

The journey of simulation has entered an era where artificial intelligence has assumed an important role. The need of this era is to build more flexible, adaptable, and autonomous tools that can take decisions in dynamic situations and also adapt to the dynamically changing environment after learning about it. Agent technology has emerged to mainly satisfy this need.

The above-mentioned demand was also present in the past, but the technology had not yet been developed to satisfy it. With the advent of new technology, intelligent applications have also become more sophisticated. Intelligent applications started their journey significantly from the application of "holons." The meaning of the word *holos* is "whole" and the meaning of the word *on* is "particle." The term "holonic" mainly refers to decentralized control in a system. The basic emphasis of a holonic system is its flexibility to adapt to the dynamic environment.

After holonic systems, competitive factors were also emphasized along with flexibility, which gave rise to the concept of "agile" systems. Agile systems mainly emphasize effective and efficient utilization of resources for prompt and flexible responses to changing customer needs. These systems are particularly effective for chaotic markets where customers get attracted to new ideas and innovative products.

With the advent of technology and frequently changing customer needs, the need arose for adaptive and autonomous decision-making tools, which gave rise to agent technology, which is mainly an idea of artificial intelligence. Agents are basically self-contained, autonomous, adaptable, active components that are capable of interacting with the physical world and the various other components of a system.

The development of a perfectly independent, self-contained, adaptive, and autonomous agent is still a challenge to both researchers over the world and practitioners in various fields of study. The emergence of new technologies is continuously shaping the approaches of designing agents. However, before turning to the various aspects of agent technology, the basic characteristics of the agents must be introduced to understand the basic definition and the concept of agents.

16.2 CHARACTERISTICS OF AGENTS

Agents are the independent, autonomous, decision-making components in a system. Therefore, there are various characteristics of agents that make them flexible and adaptive to dynamic environments. Figure 16.1 shows the characteristics of agents

FIGURE 16.1 Characteristics of the agents.

in general agreed by various authors as evident from the existing literature. Each of these characteristics is briefly explained as follows:

1. *Autonomy.* Agents should be autonomous, which means that they should not be under the direct control of outside controllers, such as users.
2. *Adaptability.* Agents should have the capability to adapt to the changing or dynamic situations arising over time.
3. *Cooperation.* Agents should have the capability to cooperate among themselves so as to achieve the end result.
4. *Proactiveness.* Agents should have the capability to take initiative for an action whenever necessary. This goal-directed behavior is actually applicable to multiagent systems.
5. *Mobility.* Agents should have the ability to change their physical locations so as to achieve the overall goal.
6. *Social ability.* Agents should have the ability to interact among themselves so as to exchange the required data. This effective social ability gives rise to cooperation among the agents.
7. *Learning.* Agents should have the ability to learn from the emerging changing environment. Learning improves agents' adaptability.
8. *Veracity.* Agents should be developed in such a way that they never communicate false information to any entity in a system.
9. *Responsiveness.* Agents must respond effectively whenever required. They must perceive what is required of them at any particular point of time and must respond accordingly.

10. *Rationality.* Agents should be rational while making decisions. They should not prevent themselves from functioning without significant reason.

All the characteristics discussed so far may not be present in all types of agents, although those that are required to be present in all agents are autonomy, adaptability, cooperation, and responsiveness. Thus, before discussing further, Section 16.3 discusses the various types of agents as proposed in the existing literature.

16.3   TYPES OF AGENTS

The classification of agents depends on several factors. The most significant among them are as follows:

1. Classification based on various application areas
2. Classification based on the agent topologies
3. Classification based on the stimulus of actions for agents

Figure 16.2 shows the classification of agents based on various factors. Each of the factors is described as follows:

Classification based on various application areas. This type is numerous and increases with the advent of agent technology and with the number of uses in various fields of study. However, the existing literature currently shows a vast number of research studies on biological agents, robotic agents, and various computational agents. These are described in brief as follows:

1. *Biological agents* may include various chemical, biological, or physical agents existing in various plant and animal lives. These may also include different bacteria living in various organisms and animals.
2. *Robotic agents* are intelligent physical agents that have sensors and various mechanical parts and are capable of performing physical movements and making independent decisions.
3. *Computational agents* are mainly software agents that are programmed to perform certain activities. The decision-making parts, the characteristics of autonomy, and the interactions among the agents and with the

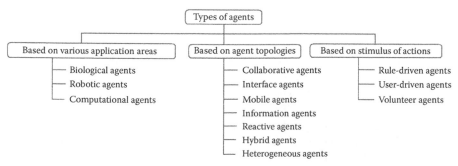

FIGURE 16.2   Types of agents.

external environment are all programmed inside the implemented agent structure.

Classification based on agent topologies. Agent topologies can be defined as the way in which a particular agent system works. The existing literature [1] has proposed the following basic agent topologies:

1. *Collaborative agents.* These agents are active in distributed environments where they interact among themselves to make a decision. Thus, the main characteristics prevailing in these types of agents are autonomy, social ability, and cooperation.

2. *Interface agents.* These types of agents interface with the users. Based on the interaction through the interface, they learn from their external environment and adapt to the requirements in their behavior. Thus, the basic characteristics prevailing in these types of agents are autonomy, learning, and adaptability.

3. *Mobile agents.* The name "mobile" for these types of agents comes from the capability to move in a network, which allows them to interact with the external environment, collect the required data, and then perform the required tasks for their users. Thus, the basic characteristics prevailing in these types of agents are mobility, learning, and responsiveness.

4. *Information agents.* These agents perform various tasks to manage large amounts of information in large networks. They can distribute, carry, and also process large amounts of information.

5. *Reactive agents.* The name "reactive" comes from the fact that these agents mainly start functioning based on external stimuli. They then return the results of their activities after reacting to the stimuli. The main characteristic prevailing in these types of agents is responsiveness.

6. *Hybrid agents.* These agents are a combination or hybridization of two or more agents' behaviors or types.

7. *Heterogeneous agents.* These agents combine the characteristics of two or more agents belonging to various agent architectures.

Classification based on the stimulus of actions for agents. Agents can also be classified based on the stimuli that they receive. These agents include rule-driven agents, user-driven agents, and volunteer agents. Each of these names indicates the stimulus that triggers the functioning of the agent. Each of these agent types is explained as follows:

1. *Rule-driven agents.* The activities of these agents are triggered by some predefined rules and assumptions. These agents start functioning whenever the assumptions are satisfied.

2. *User-driven agents.* These agents start functioning whenever a request comes from a user. The actions to be performed based on various requests are prespecified and defined beforehand in their structure. Whenever a particular request is received, then the agent starts executing particular predefined functions for that request.

3. *Volunteer agents.* These agents look for the opportunity or the requirement of the users and react automatically whenever a need arises. Thus, these agents are more autonomous and responsive in nature than the previous ones.

Before delving further into the concept of agent-based simulation, an overview of the phases of agent-based simulation needs to be provided. Thus, Section 16.4 describes the phases of agent-based simulation.

16.4 PHASES OF GENERAL AGENT-BASED SIMULATION

The phases of agent-based simulation depend on the types of agents considered. Since the types of agents vary across different applications, there is no standard number of phases for agent-based simulation. However, a generalized set of phases can be presented which is applicable to all types of agent-based simulations. The phases of general simulation are depicted in Figure 16.3.

Generally, while discussing simulation, it must be noted that agent-based simulation is typically a multiagent-based simulation, since a simulation model contains multiple agents. The phases of simulation modeling shown in Figure 16.3 are depicted in light of agent-based simulation as follows:

1. *Observation.* At first, the system that is to be simulated using agents is analyzed to identify the various agents. This stage is very important since any mistake at this stage will be carried forward to the subsequent phases and the entire agent-based simulation will be in vain.
2. *Model/theory construction.* Once the agents are properly identified, almost half of the design is done. Then, the architectures of the agents and the interactions among them are designed. Thus, the conceptual model is designed at this stage.
3. *Computational modeling.* At this stage, the rules and precedence levels that will be applied to govern the agents are decided. If it is a user-driven model, then the requests and the corresponding functionalities are defined. For volunteer agent-based models, the design for the search for possible scope for this system is completed.
4. *Software implementation.* Based on the above three stages, the package to be used is decided first and the respective required functions and modules are developed.
5. *Simulation model implementation.* After the implementation of the entire model in the software, the model is executed and the possible flaws in the

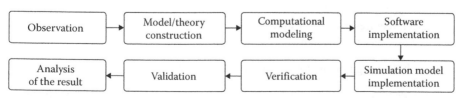

FIGURE 16.3 General phases of simulation.

model are identified. If any defect in the modeling is identified, then any of the above four stages may be repeated depending on the type of defect.

6. *Verification.* A computerized model is compared with the conceptual model to verify whether the computerized model is performing what it is supposed to perform. The input parameters and the logical structure of the implemented software model are verified at this stage.

7. *Validation.* This step is used to check whether the implemented model actually represents the real system. This process is often an iterative process in which the inconsistencies or discrepancies between the implemented model and the real system are identified, and the implemented model is improved each time a discrepancy is identified.

8. *Analysis of the result.* This last step is important since this is the reason for developing and running the simulation model. The results returned by the agents and the system as a whole are analyzed for drawing conclusions.

The above steps certainly show some differences between agent-based simulations and conventional simulations. Although agent-based simulations may not be applicable to all kinds of problems, the application of an agent-based approach has certain advantages over a conventional simulation. These advantages may include the following:

1. When agents are used, a certain extent of autonomy can be achieved in decision making.

2. Agents can be created or destroyed whenever required. This saves valuable computational resources.

3. Emergent behaviors of a system can be studied by applying multiagent-based simulation.

4. Agents can communicate among themselves, share information, learn from changing environments, adjust to new environments, and show proactive behavior. These properties make a system an intelligent one that is capable of making decisions, thereby enhancing the flexibility of the system as a whole.

16.5   DESIGN OF AGENTS

Agents are identified first and then the agent architecture is defined. Defining the architecture means designing the agent-based system to be implemented. The design of a multiagent-based simulation model can be based on the problem under study, the process to be applied, or the architecture orientation. Based on these three approaches, the design method can be classified into the following types (Figure 16.4):

1. Problem-oriented design
2. Architecture-oriented design
3. Process-oriented design

Each of these design methodologies is described as follows:

A *problem-oriented design* starts with defining the problem. Once the problem is defined, the model is based on the user's requirements. Thus, this type of designing is applicable to systems where the settings may change, which may

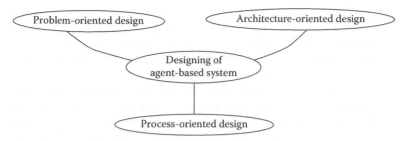

FIGURE 16.4 Types of agent-based designs.

represent the characteristics of the system. Such agents are commonly known as synthetic agents. Examples of such systems include Gaia and multiagent systems engineering (MaSE).

An *architecture-oriented design* mainly indicates a software-oriented design. This means that such a design indicates all the pros and cons of the design including the structure of the system, the details of implementation, the set of standards for governing the interactions among the agents, and the provision of the reusability features. This design actually depicts the details of all the aspects that will be required for designing a system.

A *process-oriented design* is mainly driven by the constraints of the system. The constraints may include various resource constraints. A process-oriented design may be designed in such a way that it shows dynamic behavior due to the changing states of the constraints.

In addition, all agent architectures can be classified into the following:

1. Hierarchical architecture
2. Federation architecture
3. Autonomous agent architecture (AAA)

Each of the above architectures is described as follows:

A *hierarchical architecture* has various levels of controls. The agents are divided into hierarchies of functions. Generally, the agents at the higher levels control the agents at the lower levels. Therefore, this architecture is a centralized architecture, and the flexibility of the agents is significantly reduced, especially for the lower level agents. This architecture is applicable where high degrees of control over the agents' activities are required.

A *federation architecture* divides the agents into clusters on the basis of their functions. There are three different types of federation architectures: (1) facilitator-based architecture, (2) broker-based architecture, and (3) mediator-based architecture.

1. In a *facilitator-based architecture*, the communication between the agents in a cluster is allowed easily, and the communication between the clusters is allowed through the facilitator. The facilitator acts as a gateway for communication between the clusters. The facilitator is responsible for handling the message traffic.

2. In a *broker-based architecture*, the broker acts as an intermediary between the clients and the service provider. Each broker is associated with a cluster. Whenever an external client makes a service request to a cluster agent, the broker of that cluster acts as an intermediary between the client and the cluster agents.

3. In a *mediator-based architecture*, as the name suggests, the mediator acts like a broker, but also manages the activities of the server agents. Whenever a client makes a service request, the mediator first chooses the type of agent that can best serve the client. Then, the mediator coordinates and handles all the activities related to the communication between the selected server agent and the client.

An *autonomous agent architecture* is an architecture for designing independent, intelligent agents that will work in a dynamic environment. As the name suggests, these agents are autonomous in nature and are capable of making independent decisions. The agents in this type of architecture acquire knowledge and embed the knowledge in their knowledge base. Their behavior is also goal directed. Based on the goal to achieve, a plan of action is developed and then the agents execute the developed plan.

16.6   MULTIAGENT-BASED SIMULATION IN MANUFACTURING

The combination of multiagent technology with multicriteria technology is not new in the existing literature. But the number of research publications on the combination of these approaches is not as much as the number of research studies in each of the individual categories.

The benchmark multiagent technologies are listed in Table 16.1. Among these technologies, Gaia [2] is a hierarchical agent-based architecture that uses the

TABLE 16.1
Benchmark Multiagent Technologies

Name of Multiagent Technology	Reference
Gaia	[2]
FUSION	[3]
ADEPT	[4]
ARCHON	[5]
ROADMAP	[6]
Prometheus	[7]
Tropos	[8]
PASSI	[9]
STYX	[10]
TAPAS	[11]
JADE	[12]
PROSA	[13]
ADACOR	[14]
HCBA	[15]

concepts of object-oriented analysis and design. Wooldridge et al. [2] used some concepts from FUSION [3]. Gaia is suitable for the development of systems such as ADEPT [4] and ARCHON [5]. In Gaia, every agent has a role to play and interacts with other agents in a certain predefined way based on their protocols. ROADMAP [6] is another agent-based methodology, which is an extension of Gaia for complex open systems.

The Prometheus [7] methodology includes learning of agents. It has three distinct phases of development: (1) system specification phase, in which the goals and scenarios for the system are defined; (2) architectural design phase, in which the agents' structures are defined; and (3) detailed design phase, in which the interaction protocols of the agents are defined. Another related agent-based methodology for software development is Tropos [8], which can be used in all phases of software development, starting from system analysis to the implementation phase. PASSI [9] is a multiagent system that consists of models such as system requirement models, agent society models, agent implementation models, code models, and deployment models. Some of the other significant agent-based technologies include STYX [10], TAPAS [11], and JADE [12].

Among the agent-based technologies applied in manufacturing, the holonic approaches have been well practiced. The PROSA [13] agent architecture concentrates on a hierarchical holonic system using object-oriented concepts of specialization and aggregation, whereas ADACOR [14] uses neither the hierarchical nor the decentralized system. ADACOR uses the hierarchical structure at the control level. HCBA [15] uses two holons—product and resource—to imitate their physical counterparts in the manufacturing system.

There are also numerous hybrid approaches observed in the existing literature. Some of them are listed in Table 16.2.

Barbucha [16] combined the multiagent approach with guided local search metaheuristics. Soroor et al. [17] developed an integrated supply chain coordination model by combining a fuzzy logic, an analytic hierarchical process, and a quality function deployment with the agent-based approach. Zhao et al. [18] combined system dynamics.

TABLE 16.2
Hybrid Approaches in Agent Technology

Tools Hybridized with Agent Technology	Reference
Local search meta-heuristics	[16]
Fuzzy logic, analytic hierarchy process and quality function deployment	[17]
System dynamics	[18]
Particle swarm optimization	[19]
Artificial immune algorithm	[20]
Petri net	[21]
Ant colony optimization algorithm	[22]
Simulated annealing and branch-and-bound technique	[23]
Genetic algorithm	[24]
Artificial neural network	[25]

Sinha et al. [19] combined particle swarm optimization technique with agent-based approaches. Shi and Qian [20] combined artificial immune algorithm and Hsieh [21] combined Petri net with agent-based approaches. Leung et al. [22] combined ant colony algorithm and Lee et al. [23] combined simulated annealing and branch-and-bound technique with agent-based approaches. Shao et al. [24] combined genetic algorithm, and López-Ortega and Villar-Medina [25] combined artificial neural network with agent-based approaches. Some of the other significant research studies include those by Lin et al. [26], Cheshmehgaz et al. [27], and Wang et al. [28].

Besides the benchmark agent-based technologies and methodologies, there are numerous applications of multiagent systems in manufacturing and related areas. Agent-based applications are observed in flow shop scheduling [29], shop floor scheduling for ship building yards [30], process planning and scheduling [31], freight transport [32], traveling salesman problem [33], virtual reality [34], traffic detection [35], system-on-chip design [36], recreational fishing [37], bidding optimization of wind generators [38], e-health monitoring [39], and so on. In addition, there are many research studies on the negotiation mechanisms among agents. The significant ones among these studies include the studies by Lin et al. [26], Aissani et al. [40], Vasudevan and Son [41], Bahrammirzaee et al. [42], Lin et al. [43], Shen et al. [44], Kim and Cho [45], Lee et al. [23,46], Asadzadeh and Zamanifar [47], Al-Mutawah et al. [48], Kouiss et al. [49], and Miyashita [50].

16.7 SOME MULTIAGENT MODELS

In this section, some of the most practiced multiagent models are discussed: Gaia, Prometheus, and MaSE.

16.7.1 GAIA METHODOLOGY

Gaia is a hierarchical agent-based methodology. Its characteristics are given as follows:

1. It uses cooperative and autonomous agents.
2. It supports both macro-level (inter-agent) and micro-level (intra-agent) aspects.
3. It can handle heterogeneous agents, independent of programming languages and agent architectures.
4. It emphasizes a coarse-grained computational system.
5. It assigns a specific role to each agent.
6. Its methodology consists of six models—agent model, role model, interaction model, services model, organizational model, and environmental model.
7. It is divided distinctly into analysis and design phases.
   a. The analysis phase defines the overall structure of the agent system. This phase also defines the preliminary interaction model using a protocol. Each protocol is defined by a name, an initiator role, a respondent role (partner), inputs, outputs, and the description of the protocol.

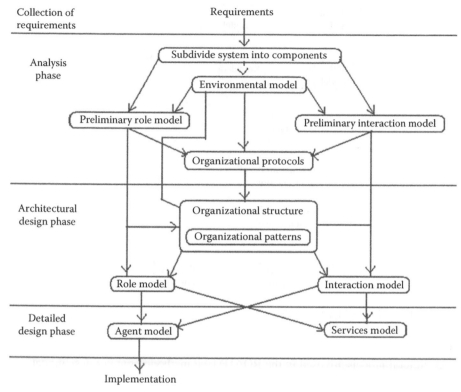

FIGURE 16.5 Gaia methodology.

    b. The design phase is further divided into architectural design and detailed design phases.

       i. The architectural design phase defines the overall organizational structure, role models, and interaction models.

      ii. The detailed design phase defines the detailed design of each agent and its services.

  8. It is appropriate for closed distributed systems where the constituent components are known at design time.

The overall Gaia methodology is depicted in Figure 16.5.

16.7.2 ROADMAP METHODOLOGY

ROADMAP is an extension of the Gaia methodology so that the methodology can be applied to open the system environment. Its specific characteristics are described as follows:

  1. It is an extension of Gaia by the addition of use cases.

  2. It is applicable to open system environments as well as closed system environments.

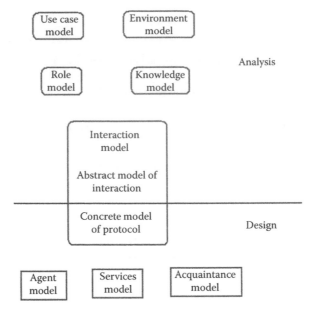

FIGURE 16.6   ROADMAP methodology.

3. Several models are used in the ROADMAP methodology: use case model, environment model, knowledge model, role model, protocol model, and acquaintance model.
   a. The use case model is used for identifying the requirements of the system.
   b. The environment model is derived from the use case model, and it describes the system environment.
   c. The knowledge model is derived from the environment model, and it describes the domain knowledge of the system.
   d. The role model defines the roles of the agents. Each role has four attributes—responsibility, permission, activity, and protocol.
   e. The interaction model (also called the protocol model) describes the relationship among agents. The protocol model is an addition by ROADMAP over Gaia model.
   f. The acquaintance model defines the communication among agents.

Figure 16.6 gives an overview of the ROADMAP methodology.

16.7.3   PROMETHEUS METHODOLOGY

Prometheus is another agent-based methodology that has ample industrial and academic applications. This methodology has three phases—system specification phase, architectural design phase, and detailed design phase.

The specific characteristics of each phase of this methodology are provided as follows:

1. The system specification phase identifies the system goals, defines the agents' interfaces with the environment in terms of actions and percepts, and prepares functionality schema. The functionality schema consists of name, description, inputs (percepts), outputs (actions), and interactions. The output of the system specification phase serves as the input to the architectural design phase.
2. The architectural design phase describes the overall structure of the system. It basically chooses the agents for the system and defines the interactions among them. The specific diagrams used in the architectural design phase include a data coupling diagram and an agent acquaintance diagram.
3. The detailed design phase describes in detail about how each agent works. The specific diagrams used in this phase include an agent overview diagram and a process diagram. In the detailed design phase, the details of each plan, event, and data are described at the fundamental level. It supports the entire design and development of a system from start to end.

Figure 16.7 shows a schematic diagram of the Prometheus methodology.

16.7.4 PASSI Methodology

PASSI is an agent society model. The details of this model as well as its characteristics are described as follows:

1. It is developed step by step depending on the code requirement.
2. It uses object-oriented software engineering.
3. It uses UML notations to define the logic.
4. In this model, an activity diagram is used to model the plans.
5. Goals in this model are considered as nonfunctional requirements.
6. This methodology consists of five models—system requirement model, agent society model, agent implementation model, code model, and deployment model.
   a. The system requirement model consists of four phases—domain description phase, agent identification phase, role identification phase, and task specification phase.
      i. The result of the domain description phase is the use case diagram that makes use of the sequence diagram.
      ii. The agent identification phase identifies the functionalities of the agents that are represented by the activity diagrams.
      iii. The role identification phase defines the roles of the agents and the relationships among them.
      iv. The task specification phase subdivides all the tasks among the ordinary agents and the interacting agents.

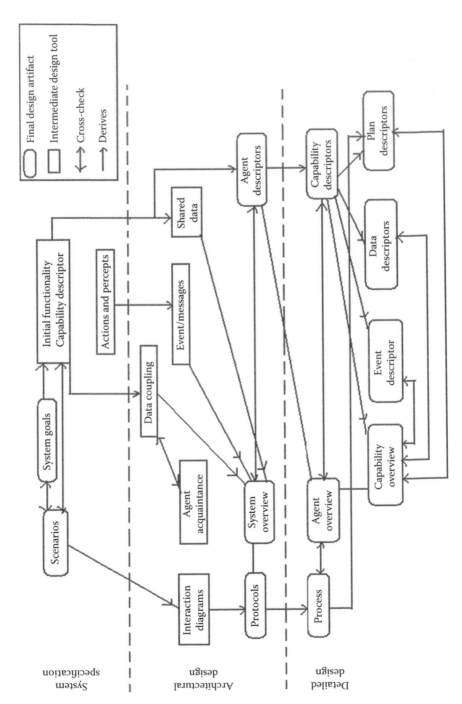

FIGURE 16.7 Prometheus methodology.

b. The agent society model has three functions—ontology description, role description, and protocol description. The ontology description describes the structure of the agents' organization; the role description defines the life of the agents based on their roles; and the protocol description defines the protocols to be executed by the agents.

c. The agent implementation model is responsible for defining the behavior and structure of the agents. Thus, the attributes of the agents, the methods to be executed by the agents, and the communications among the agents are defined. This facilitates the development of the codes for the agents.

d. There are two aspects of the code model—reuse of codes and production of codes. Codes are written in such a way that they can be reused. Code reuse makes the development of a system faster.

e. The deployment model is especially used to define the behavior of mobile agents, since mobile agents need special attention in terms of their requirements related to software environment.

Figure 16.8 shows the overall functioning and structure of the PASSI methodology.

16.7.5 MaSE Methodology

MaSE is an improvement over Gaia. Its specific characteristics are as follows:

1. It provides specific tools for automatic code generation.
2. It supports the entire software development life cycle.
3. It supports heterogeneous multiagent systems independent of any particular agent architecture.
4. It uses the concepts of a unified modeling language (UML) and a rational unified process (RUP).
5. It defines goals, agent types, and communication among agents through graphic-based models.
6. It is divided into two phases—analysis phase and design phase.
   a. In the analysis phase, the system goals are derived from the initial requirements. The system goals define the system requirements. Use cases are drawn from the system requirements. Next, the agents' roles are decided, and communication messages between the roles are decided and represented by sequence diagrams.
   b. In the design phase, agents are created, communications among them are established, agent classes are created, and the design of the entire system is done. The system design uses deployment diagrams to represent the system structure, the number of agents to be used, and the actual locations where the agents will be used.

Figure 16.9 shows the overall activities of MaSE.

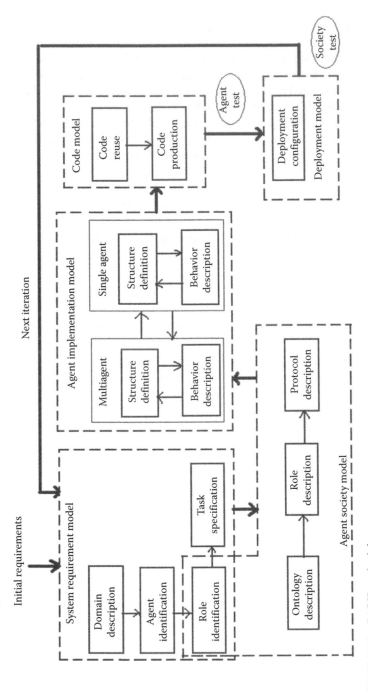

FIGURE 16.8   PASSI methodology.

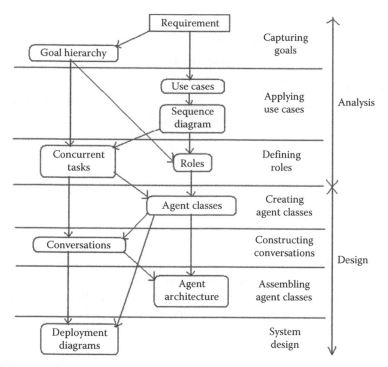

FIGURE 16.9 MaSE methodology.

16.7.6 TROPOS METHODOLOGY

Tropos is actually a software development methodology with agent orientation. Its specific characteristics are as follows:

1. It makes use of UML and provides applications for the development of beliefs, desire, and intentions (BDI) agents.
2. It has five phases—early requirement phase, late requirement phase, architectural design phase, detailed design phase, and implementation phase.
   a. The early requirement phase identifies the stakeholders or actors and also identifies their objectives or goals.
   b. The late requirement phase establishes the relationships among the actors.
   c. The architectural design phase breaks the goals into subgoals and identifies the respective actors.
   d. The detailed design phase defines the actors and the communications and coordinations among them.
   e. The implementation phase makes the actual programming for the agent-based software as designed in the previous phases.

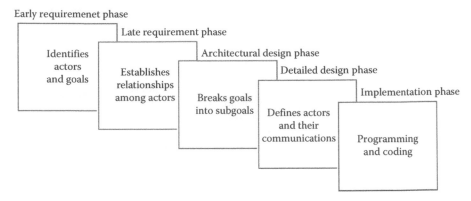

Early requiremenet phase

Late requirement phase

Identifies actors and goals

Architectural design phase

Establishes relationships among actors

Detailed design phase

Breaks goals into subgoals

Implementation phase

Defines actors and their communications

Programming and coding

FIGURE 16.10   Tropos methodology.

3. In this methodology, agents, goals, and communications among the agents are used for the early and late requirement phases.
4. Actor diagrams are used to represent the actors and the relationships among them. The other diagrams used in Tropos are activity diagrams and UML sequence diagrams.

Figure 16.10 shows a schematic diagram of the phases of the Tropos methodology.

16.8   APPLICATIONS OF AGENT-BASED SIMULATION

There is a large number of application areas where agent-based simulation has been used. It can easily be foreseen that the number of agent-based application areas will increase more with the advent of technology.

Hare and Deadman [51] provide an overview of the applications of agent-based simulation for environmental problems. The range of application areas in environmental modeling where agent-based simulation has been applied is also mentioned in this research study. Some of these areas are as follows:

1. Rural water resource management (Bali model, SHADOC model, CATCHSCAPE model, Lake model)
2. Urban water demand management (Thames model)
3. Flood mitigation decision support (MAGIC model)
4. Animal waste management (Biomass model)
5. Rangeland resources management (Rangeland model)
6. Agricultural land-use change models (FEARLUS model, LUCITA model)
7. Recreation management (Grand Canyon model)

In addition, there are multiagent-based simulation applications in safety-critical systems [52], online trading [53], traffic flow of construction equipment [54], strategic applications in wholesale electricity markets [55], behavioral studies for employees in an organization [56], electron optical systems [57], labor market applications [58], medical systems [59], urban management systems [60], agricultural applications [61], banking sectors [62], consumer affordability [63], purchasing behaviors of consumers [64], emergency evacuation system with fire [65], social networks [66], inventory systems [67], traffic signal control systems [18], mobile networks [68], and numerous others.

16.9 CONCLUSION

This chapter has given a glimpse into the world of agent-based simulation and has worked as a window that has shown the various aspects of agent-based simulation. Besides providing a discussion of the characteristics, types, and design of agents and agent-based systems, this chapter has also provided a useful depiction of some of the famous agent-based systems. In Section 16.8, the applications of agent-based simulation have been presented. Thus, this chapter is expected to help the reader to get an overview of agent-based simulations, their various applications, and some famous agent-based simulation models.

REFERENCES

1. Paolucci, M. and Sacile, R. (2005). *Agent-Based Manufacturing and Control Systems: New Agile Manufacturing Solutions for Achieving Peak Performance.* New York: CRC Press.
2. Wooldridge, M., Jennings, N. R., and Kinny, D. (2000). The Gaia methodology for agent-oriented analysis and design. *Autonomous Agents and Multi-Agent Systems* 3(3): 285–312.
3. Coleman, D., Arnold, P., Bodoff, S., Dollin, C., Gilchrist, H., Hayes, F., and Jeremaes, P. (1994). *Object-Oriented Development: The FUSION Method.* Hemel Hempstead: Prentice Hall International.
4. Jennings, N. R., Faratin, P., Johnson, M. J., Norman, T. J., O'Brien, P., and Wiegand, M. E. (1996). Agent-based business process management. *International Journal of Cooperative Information Systems* 5(2/3): 105–130.
5. Jennings, N. R. (1996). Using ARCHON to develop real-world DAI applications. 1. *IEEE Expert* 11(6): 64–70.
6. Juan, T., Pearce, A., and Sterling, L. (2002). ROADMAP: Extending the Gaia methodology for complex open systems. In M. Gini, T. Ishida, C. Castelfranchi, and W. L. Johnson (Eds.), *Proceedings of the 1st International Joint Conference on Autonomous Agents and Multiagent Systems*, Bologna, Italy. New York: ACM Press, July 15–19, pp. 3–10.
7. Padgham, L. and Winikoff, M. (2004). *Developing Intelligent Agent Systems—A Practical Guide.* Chichester: Wiley.
8. Bresciani, P., Giorgini, P., Giunchiglia, F., Mylopoulos, J., and Perini, A. (2004). Tropos: An agent-oriented software development methodology. *Journal of Autonomous Agents and Software Development Methodologies* 8: 203–236.
9. Burrafato, P. and Cossentino, M. (2002). Designing a multi-agent solution for a bookstore with the PASSI methodology. *Proceedings of the 4th International Bi-Conference Workshop on Agent-Oriented Information Systems*, Toronto, ON. http://mozart.csai. unipa.it/passi.
10. Geoff, B., Stephen, C., and Martin, P. (2001). The Styx agent methodology. The Information Science Discussion Paper Series 2001/02, Department of Information Science, University of Otago, New Zealand. http://divcom.otago.ac.nz/infosci.

11. Holmgren, J., Davidsson, P., Persson, J. A., and Ramstedt, L. (2012). TAPAS: A multi-agent-based model for simulation of transport chains. *Simulation Modelling Practice and Theory* 23: 1–18.
12. Kanaga, E. G. M. and Valarmathi, M. L. (2012). Multi-agent based patient scheduling using particle swarm optimization. *Procedia Engineering* 30: 386–393.
13. van Brussel, H., Wyns, J., Valckenaers, P., and Bongaerts, L. (1998). Reference architecture for holonic manufacturing systems: PROSA. *Computers in Industry* 37(3): 255–274.
14. Leitão, P., Colombo, A., and Restivo, F. (2005). ADACOR: A collaborative production automation and control architecture. *IEEE Intelligent Systems* 20(1): 58–66.
15. Chirn, J. and McFarlane, D. (2000). A component-based approach to the holonic control of a robot assembly cell. *Proceedings of the IEEE 17th International Conference on Robotics and Automation*, April 24–28, San Francisco, CA.
16. Barbucha, D. (2012). Agent-based guided local search. *Expert Systems with Applications* 39: 12032–12045.
17. Soroor, J., Tarokh, M. J., Khoshalhan, F., and Sajjadi, S. (2012). Intelligent evaluation of supplier bids using a hybrid technique in distributed supply chains. *Journal of Manufacturing Systems* 31(2): 240–252.
18. Zhao, J., Mazhari, E., Celik, N., and Son, Y.-J. (2011). Hybrid agent-based simulation for policy evaluation of solar power generation systems. *Simulation Modelling Practice and Theory* 19: 2189–2205.
19. Sinha, A. K., Aditya, H. K., Tiwari, M. K., and Chan, F. T. S. (2011). Agent oriented petroleum supply chain coordination: Co-evolutionary Particle Swarm Optimization based approach. *Expert Systems with Applications* 38(5): 6132–6145.
20. Shi, X. and Qian, F. (2011). A multi-agent immune network algorithm and its application to murphree efficiency determination for the distillation column. *Journal of Bionic Engineering* 8(2): 181–190.
21. Hsieh, F.-S. (2010). Design of reconfiguration mechanism for holonic manufacturing systems based on formal models. *Engineering Applications of Artificial Intelligence* 23(7): 1187–1199.
22. Leung, C. W., Wong, T. N., Mak, K. L., and Fung, R. Y. K. (2010). Integrated process planning and scheduling by an agent-based ant colony optimization. *Computers & Industrial Engineering* 59(1): 166–180.
23. Lee, W.-C., Chen, S.-K., and Wu, C.-C. (2010). Branch-and-bound and simulated annealing algorithms for a two-agent scheduling problem. *Expert Systems with Applications* 37(9): 6594–6601.
24. Shao, X., Li, X., Gao, L., and Zhang, C. (2009). Integration of process planning and scheduling—A modified genetic algorithm-based approach. *Computers & Operations Research* 36(6): 2082–2096.
25. López-Ortega, O. and Villar-Medina, I. (2009). A multi-agent system to construct production orders by employing an expert system and a neural network. *Expert Systems with Applications* 36(2): 2937–2946.
26. Lin, Y.-I., Tien, K.-W., and Chu, C.-H. (2012). Multi-agent hierarchical negotiation based on augmented price schedules decomposition for distributed design. *Computers in Industry* 63(6): 597–609.
27. Cheshmehgaz, H. R., Desa, M. I., and Wibowo, A. (2013). A flexible three-level logistic network design considering cost and time criteria with a multi-objective evolutionary algorithm. *Journal of Intelligent Manufacturing* 24: 277–293.
28. Wang, L., Tang, D.-B., Gu, W.-B., Zheng, K., Yuan, W.-D., and Tang, D.-S. (2012). Pheromone-based coordination for manufacturing system control. *Journal of Intelligent Manufacturing* 23(3): 747–757.

29. Gómez-Gasquet, P., Andrés, C., and Lario, F.-C. (2012). An agent-based genetic algorithm for hybrid flowshops with sequence dependent setup times to minimise makespan. *Expert Systems with Applications* 39(9): 8095–8107.

30. Choi, H. S. and Park, K. H. (1997). Shop-floor scheduling at shipbuilding yards using the multiple intelligent agent system. *Journal of Intelligent Manufacturing* 8(6): 505–515.

31. Li, X., Gao, L., and Shao, X. (2012). Anactive learning genetic algorithm for integrated process planning and scheduling. *Expert Systems with Applications* 39(8): 6683–6691.

32. Anand, N., Yang, M., van Duin, J. H. R., and Tavasszy, L. (2012). GenCLOn: An ontology for city logistics. *Expert Systems with Applications* 39(15): 11944–11960.

33. Ilie, S. and Bădică, C. (2011). Multi-agent approach to distributed ant colony optimization. *Science of Computer Programming* 78: 762–774.

34. Mahdjoub, M., Monticolo, D., Gomes, S., and Sagot, J.-C. (2010). A collaborative design for usability approach supported by virtual reality and a multi-agent system embedded in a PLM environment. *Computer-Aided Design* 42(5): 402–413.

35. Chen, B., Cheng, H. H., and Palen, J. (2009). Integrating mobile agent technology with multi-agent systems for distributed traffic detection and management systems. *Transportation Research Part C* 17(1): 1–10.

36. Trappey, C. V., Trappey, A. J. C., Huang, C.-J., and Ku, C. C. (2009). The design of a JADE-based autonomous workflow management system for collaborative SoC design. *Expert Systems with Applications* 36(2): 2659–2669.

37. Gao, L. and Hailu, A. (2012). Ranking management strategies with complex outcomes: An AHP-fuzzy evaluation of recreational fishing using an integrated agent-based model of a coral reef ecosystem. *Environmental Modelling & Software* 31: 3–18.

38. Li, G. and Shi, J. (2012). Agent-based modeling for trading wind power with uncertainty in the day-ahead wholesale electricity markets of single-sided auctions. *Applied Energy* 99: 13–22.

39. Su, C. J. (2008). Mobile multi-agent based, distributed information platform (MADIP) for wide-area e-health monitoring. *Computers in Industry* 59(1): 55–68.

40. Aissani, N., Bekrar, A., Trentesaux, D., and Beldjilali, B. (2012). Dynamic scheduling for multi-site companies: A decisional approach based on reinforcement multi-agent learning. *Journal of Intelligent Manufacturing* 23(6): 2513–2529.

41. Vasudevan, K. and Son, Y.-J. (2011). Concurrent consideration of evacuation safety and productivity in manufacturing facility planning using multi-paradigm simulations. *Computers & Industrial Engineering* 61(4): 1135–1148.

42. Bahrammirzaee, A., Chohra, A., and Madani, K. (2011). An artificial negotiating agent modeling approach embedding dynamic offer generating and cognitive layer. *Neurocomputing* 74(16): 2698–2709.

43. Lin, Y.-I., Chou, Y.-W., Shiau, J.-Y., and Chu, C.-H. (2013). Multi-agent negotiation based on price schedules algorithm for distributed collaborative design. *Journal of Intelligent Manufacturing* 24: 545–557.

44. Shen, W., Maturana, F., and Norrie, D. H. (2000). Enhancing the performance of an agent-based manufacturing system through learning and forecasting. *Journal of Intelligent Manufacturing* 11(4): 365–380.

45. Kim, H. S. and Cho, J. H. (2010). Supply chain formation using agent negotiation. *Decision Support Systems* 49(1): 77–90.

46. Lee, Y. H., Kumara, S. R. T., and Chatterjee, K. (2003). Multi-agent-based dynamic resource scheduling for distributed multiple projects using a market mechanism. *Journal of Intelligent Manufacturing* 14(5): 471–484.

47. Asadzadeh, L. and Zamanifar, K. (2010). An agent-based parallel approach for the job shop scheduling problem with genetic algorithms. *Mathematical and Computer Modelling* 52(11/12): 1957–1965.

48. Al-Mutawah, K., Lee, V., and Cheung, Y. (2009). A new multi-agent system framework for tacit knowledge management in manufacturing supply chains. *Journal of Intelligent Manufacturing* 20(5): 593–610.
49. Kouiss, K., Pierreval, H., and Mebarki, N. (1997). Using multi-agent architecture in FMS for dynamic scheduling. *Journal of Intelligent Manufacturing* 8(1): 41–47.
50. Miyashita, K. (1998). CAMPS: A constraint-based architecture for multiagent planning and scheduling. *Journal of Intelligent Manufacturing* 9(2): 147–154.
51. Hare, M. and Deadman, P. (2004). Further towards a taxonomy of agent-based simulation models in environmental management. *Mathematics and Computers in Simulation* 64: 25–40.
52. Yujun, Z., Zhongwei, X., and Meng, M. (2012). Agent-based simulation framework for safety critical system. 2012 International Workshop on Information and Electronics Engineering (IWIEE). *Procedia Engineering* 29: 1060–1065.
53. Zhong, J., Zhu, W., Wu, Y., and Wang, K. (2012). Agent-based simulation of online trading. *Systems Engineering Procedia* 5: 437–444.
54. Kim, K. and Kim, K. J. (2010). Multi-agent-based simulation system for construction operations with congested flows. *Automation in Construction* 19: 867–874.
55. Azadeh, A., Skandari, M. R., and Maleki-Shoja, B. (2010). An integrated ant colony optimization approach to compare strategies of clearing market in electricity markets: Agent-based simulation. *Energy Policy* 38: 6307–6319.
56. Wang, J., Gwebu, K., Shanker, M., and Troutt, M. D. (2009). An application of agent-based simulation to knowledge sharing. *Decision Support Systems* 46: 532–541.
57. Szilagyi, M. N. (2013). Determination of the potential distribution of electron optical systems by agent-based simulation. *Optik.* http://dx.doi.org/10.1016/j.ijleo.2012.10.026.
58. Chaturvedi, A., Mehta, S., Dolk, D., and Ayer, R. (2005). Agent-based simulation for computational experimentation: Developing an artificial labor market. *European Journal of Operational Research* 166: 694–716.
59. Perez, P., Dray, A., Moore, D., Dietze, P., Bammer, G., Jenkinson, R., Siokou, C., Green, R., Hudson, S. L., and Maher, L. (2012). SimAmph: An agent-based simulation model for exploring the use of psychostimulants and related harm amongst young Australians. *International Journal of Drug Policy* 23: 62–71.
60. Gao, L., Durnota, B., Ding, Y., and Dai, H. (2012). An agent-based simulation system for evaluating gridding urban management strategies. *Knowledge-Based Systems* 26: 174–184.
61. Diao, J., Zhu, K., and Gao, Y. (2011). Agent-based simulation of durables dynamic pricing. *Systems Engineering Procedia* 2: 205–212.
62. Zhang, T. and Zhang, D. (2007). Agent-based simulation of consumer purchase decision-making and the decoy effect. *Journal of Business Research* 60: 912–922.
63. Peizhong, Y., Xin, W., and Tao, L. (2011). Agent-based simulation of fire emergency evacuation with fire and human interaction model. *Safety Science* 49: 1130–1141.
64. Ronald, N., Dignum, V., Jonker, C., Arentze, T., and Timmermans, H. (2012). On the engineering of agent-based simulations of social activities with social networks. *Information and Software Technology* 54: 625–638.
65. Fu-gui, D., Hui-mei, L., and Bing-de, L. (2012). Agent-based simulation model of single point inventory system. *Systems Engineering Procedia* 4: 298–304.
66. Fang, F. C., Xu, W. L., Lin, K. C., Alam, F., and Potgieter, J. (2013). Matsuoka neuronal oscillator for traffic signal control using agent-based simulation. *Procedia Computer Science* 19: 389–395.
67. Balke, T., De Vos, M., and Padget, J. (2011). Analysing energy-incentivized cooperation in next generation mobile networks using normative frameworks and an agent-based simulation. *Future Generation Computer Systems* 27: 1092–1102.
68. Feldman, T. (2011). Leverage regulation: An agent-based simulation. *Journal of Economics and Business* 63: 431–440.

17 Continuous System Simulation

17.1 INTRODUCTION

Continuous system simulation is applicable to systems where the variables are continuous in nature. Thus, such systems are generally defined by ordinary differential equations or sets of differential and algebraic equations. There are many ways of implementing such systems. A significant number of software packages are available in the market for implementing such continuous systems.

There are several aspects of continuous system simulation. Thus, the various sections in this chapter discuss the related aspects. Section 17.2 presents a brief discussion of the various approaches to continuous system simulations. These approaches include various ordinary differential equation approaches. Section 17.3 provides brief concepts of the various integration methods used in various software packages in continuous system simulation languages (CSSLs). The methods that are frequently applied are the Euler method, the Adams predictor–corrector method, and the Runge–Kutta method. Section 17.4 discusses the various validation schemes for the models that are built for continuous system simulations. These validation schemes are divided into internal and external validation schemes. Section 17.5 shows the various application areas of continuous system simulations that have been dealt with up to now. Section 17.6 briefly describes the evolution of CSSLs. Sections 17.7 and 17.8 mention the features and types of CSSLs, respectively. Section 17.9 briefly describes the features of some CSSLs. Section 17.10 provides the conclusion of this chapter.

17.2 APPROACHES TO CONTINUOUS SYSTEM SIMULATION

The various mathematical concepts that are used to represent or model continuous systems include ordinary differential equations, partial differential equations, and differential algebraic equations. The main approach for solving the partial differential equations and the differential algebraic equations is to convert them into ordinary differential equations that can then be solved by a solver for ordinary differential equations [1]. In this chapter, we discuss only the ordinary and partial differential equations in brief.

17.2.1 ORDINARY DIFFERENTIAL EQUATIONS

The basic equation on which we apply various approximations and thus various algorithms is

$$\frac{\mathrm{d}x}{\mathrm{d}t} = f(x(t), u(t), t) \tag{17.1}$$

where:

    $x(t)$ is the state of variable x at time t

    $u(t)$ represents the value input vector at time t and the initial value of x is given by

$$x(t = 0) = x_0 \tag{17.2}$$

Such an equation can be expanded by a Taylor series. After expansion, the resulting expression can be approximated by various methods. The extent of approximation determines the accuracy of a particular approximation method. However, by applying the Taylor series expansion, we get the following expression:

$$x(t + h) = x(t) + \frac{\mathrm{d}x(t)}{\mathrm{d}t} h + \frac{\mathrm{d}^2 x(t)}{\mathrm{d}t^2} \frac{h^2}{2!} + \cdots \tag{17.3}$$

If x_i represents the ith state, then the above expression will become

$$x_i(t + h) = x_i(t) + \frac{\mathrm{d}x_i(t)}{\mathrm{d}t} h + \frac{\mathrm{d}^2 x_i(t)}{\mathrm{d}t^2} \frac{h^2}{2!} + \cdots \tag{17.4}$$

where:

    h is the step size

If we assume $n + 1$ terms of the Taylor series, the order of second derivative $\mathrm{d}^2 x_i(t)/\mathrm{d}t^2$ will be $n - 2$ since the second derivative is multiplied by h^2. Similarly, the order of the third derivative will be $n - 3$ since it will be multiplied by h^3, and so on. Thus, the approximation may be done up to the derivative multiplied by h^n where n is known as the approximation order, and the integration method to be applied is said to be of the order of n.

Different methods take different numbers of terms of the expanded series, and thus truncation occurs. Because of truncation, the method applied introduces an error that is called the truncation error. But with the increase of the power of h, the value of the terms decreases, and therefore the truncation error is approximated by h^{n+1}. Naturally, the accuracy of a method increases with the increase of approximation order. Section 17.3 discusses the extent of approximation by various methods of integration. But before that discussion, a brief introduction to partial differential equations is presented in Section 17.2.2.

17.2.2 PARTIAL DIFFERENTIAL EQUATIONS

Modeling with partial differential equations is much more difficult than that with ordinary differential equations. In most cases, a set of three to four coupled partial differential equations is obtained which are required to be handled together. Therefore, the numerical solutions of these equations are even more difficult. However, the development of various software packages on partial differential equations has made the modeling easier.

There are various methods to deal with s sets of partial differential equations, and the most popular among them is the method-of-lines (MOL) approach, which converts a set of partial differential equations into a set of ordinary differential equations that can then be solved by a solver.

The main idea of the MOL method is to identify some discrete points in a solution space for a variable rather than searching the entire solution space. Let there be two independent variables x and t, and a dependent variable f that can be represented as $f(x,t)$. Now consider the following partial differential equation:

$$\frac{\partial f}{\partial t} = \mu \frac{\partial^2 f}{\partial x^2} \tag{17.5}$$

Thus, instead of looking for a solution of f over a two-dimensional (2D) space for variables x and t, we will search for some discrete points, and thus we discretize the entire solution space. In expression (17.5), if we replace the second-order expression, $\partial^2 f/\partial x^2$, then the resulting equation can be solved easily by a standard solver for ordinary differential equations. For example, we replace $\partial^2 f/\partial x^2$ at $x = x_i$ by the following finite difference expression:

$$\frac{\partial^2 f}{\partial x^2} \quad \text{at} \quad (x = x_i) = \frac{f_{i+1} - 2f_i + f_{i-1}}{\delta x^2} \tag{17.6}$$

Substituting the above expression in Equation 17.5, we have

$$\frac{\partial f}{\partial t} = \mu \left(\frac{f_{i+1} - 2f_i + f_{i-1}}{\delta x^2} \right) \tag{17.7}$$

This expression can now be solved by a standard ordinary differential equation solver. Here, δx represents the distance between the two discrete neighboring points in the solution space.

17.3   INTEGRATION METHODS

Section 17.2 has mentioned that the approximation of ordinary differential equations is done by various integration methods. In this section, a brief description of some of these integration methods is presented. The integration methods that are used frequently by various software packages are as follows:

1. Euler method
2. Adams predictor–corrector method
3. Runge–Kutta method

Each of these three integration methods is briefly described in Sections 17.3.1 through 17.3.3.

17.3.1 EULER METHOD

The Euler method is the simplest integration method. This method approximates the Taylor series expansion by truncating the expression up to the first-order term, which results in the following expression:

$$x_i(t + h) = x_i(t) + \frac{dx_i(t)}{dt}h \tag{17.8}$$

The respective vector form of the above equation can be expressed as follows:

$$x(t + h) = x(t) + f(x(t),t)h \tag{17.9}$$

There are two approaches to the Euler algorithm— the forward Euler algorithm and the backward Euler algorithm. The steps of the forward Euler algorithm [assuming $x(t = t_0) = x_0$] are as follows:

Step 1:

$$\left.\frac{dx}{dt}\right|_{t=t_0} = f(x(t_0),t_0)$$

$$x(t_0 + h) = x(t_0) + h\left.\frac{dx}{dt}\right|_{t=t_0}$$

Step 2:

$$\left.\frac{dx}{dt}\right|_{t=t_0+h} = f(x(t_0 + h),t_0 + h)$$

$$x(t_0 + 2h) = x(t_0 + h) + h\left.\frac{dx}{dt}\right|_{t=t_0+h}$$

Step 3:

$$\left.\frac{dx}{dt}\right|_{t=t_0+2h} = f(x(t_0 + 2h),t_0 + 2h)$$

$$x(t_0 + 3h) = x(t_0 + 2h) + h\left.\frac{dx}{dt}\right|_{t=t_0+2h}$$

and so on.

The backward Euler algorithm uses expression (17.10) as the approximation algorithm. The iterations will be similar as shown in the case of the forward Euler algorithm.

$$x(t + h) = x(t) + f(x(t + h),t + h)h \tag{17.10}$$

17.3.2 PREDICTOR–CORRECTOR METHOD

The general predictor–corrector method starts with the expression for the forward Euler method that serves as the predictor, and we use the result of this method as the expression for the backward Euler method. We iterate on this backward Euler step. Thus, the steps of the predictor–corrector method are as follows:

Predictor:

$$\frac{dx}{dt}\bigg|_{t=t_k} = f(x(t_k),t_k)$$

$$x^{P}(t_{k+1}) = x(t_k) + h\frac{dx}{dt}\bigg|_{t=t_k}$$

Corrector 1:

$$\left(\frac{dx}{dt}\right)^{P}\bigg|_{t=t_{k+1}} = f(x^{P}(t_{k+1}),t_{k+1})$$

$$x^{P}(t_{k+1}) = x(t_k) + h\left(\frac{dx}{dt}\right)^{P}\bigg|_{t=t_{k+1}}$$

Corrector 2:

$$\left(\frac{dx}{dt}\right)^{C1}\bigg|_{t=t_{k+1}} = f(x^{C1}(t_{k+1}),t_{k+1})$$

$$x^{C2}(t_{k+1}) = x(t_k) + h\left(\frac{dx}{dt}\right)^{C1}\bigg|_{t=t_{k+1}}$$

Corrector 3:

$$\left(\frac{dx}{dt}\right)^{C2}\bigg|_{t=t_{k+1}} = f\left(x^{C2}(t_{k+1}),t_{k+1}\right)$$

$$x^{C3}(t_{k+1}) = x(t_k) + h\left(\frac{dx}{dt}\right)^{C2}\bigg|_{t=t_{k+1}}$$

and so on.

This type of iteration is called a fixed point iteration. There are also various combinations of predictor–corrector methods, each of which is an improvement over the others for some aspects of correctness of solutions.

17.3.3 RUNGE–KUTTA METHOD

The Runge–Kutta method starts with the predictor–corrector model. The predictor–corrector model proposed by Heun's method is given by

Predictor:

$$\left. \frac{dx}{dt} \right|_{t=t_k} = f(x_k, t_k)$$

$$x^P = x(t_k) + h\,\beta_{11} \left. \left(\frac{dx}{dt} \right) \right|_{t=t_k}$$

Corrector:

$$\left(\frac{dx}{dt} \right)^P = f(x^P, t_k + \alpha_1\, h)$$

$$x^P(t_{k+1}) = x(t_k) + h \left[\beta_{21} \left. \left(\frac{dx}{dt} \right) \right|_{t=t_k} + \beta_{22} \left(\frac{dx}{dt} \right)^P \right]$$

where:
   β_{ij} are the weighting factors for the various state derivatives
   α_1 is the time step in which the first stage is evaluated

Applying these equations in a Taylor series, we get

$$x^C(t_{k+1}) = x(t_k) + h(\beta_{21} + \beta_{22})f_k + \frac{h^2}{2} \left(2\,\beta_{11}\,\beta_{22}\, \frac{\partial f_k}{\partial x}\, f_k + 2\,\alpha_1\,\beta_{22}\, \frac{\partial f_k}{\partial t} \right) \quad (17.11)$$

The Taylor series of $x(t_{k+1})$, after truncation, becomes

$$x(t_{k+1}) = x(t_k) + h\,f_k + \frac{h^2}{2} \left(\frac{\partial f_k}{\partial x}\, f_k + \frac{\partial f_k}{\partial t} \right) \quad (17.12)$$

Comparing expression (17.11) with (17.12), we get

$$\beta_{21} + \beta_{22} = 1$$
$$2\,\beta_{11}\,\beta_{22} = 1 \quad (17.13)$$
$$2\,\alpha_1\,\beta_{22} = 1$$

If Heun's method is applied, then that method is characterized by

$$\alpha = \begin{bmatrix} 1 \\ 1 \end{bmatrix} \quad \text{and} \quad \beta = \begin{bmatrix} 1 & 0 \\ 0.5 & 0.5 \end{bmatrix} \quad (17.14)$$

If Butcher's method is applied, then we have

$$\alpha = \begin{bmatrix} 0.5 \\ 1 \end{bmatrix} \quad \text{and} \quad \beta = \begin{bmatrix} 0.5 & 0 \\ 0 & 1 \end{bmatrix} \tag{17.15}$$

and its corresponding table is as follows:

0	0	0
½	½	0
x	0	1

Using this method, the predictor–corrector model becomes

Predictor:

$$\left. \frac{dx}{dt} \right|_{t=t_k} = f(x_k, t_k)$$

$$x^P(t_{k+1/2}) = x(t_k) + \frac{h}{2} \left(\frac{dx}{dt} \right) \Big|_{t=t_k}$$

Corrector:

$$\left(\frac{dx}{dt} \right)^P \Big|_{t=t_{k+1/2}} = f(x^P(t_{k+1/2}), t_{k+1/2})$$

$$x^C(t_{k+1}) = x(t_k) + h \left(\frac{dx}{dt} \right)^P \Big|_{t=t_{k+1/2}}$$

The entire method is called the Runge–Kutta method.

17.4 VALIDATION SCHEMES

After building a continuous system simulation model, the next step is to validate the model. Validation, in general, can be defined as a way to check whether a system meets the specified requirements.

Various types of validation schemes are shown in Figure 17.1. Validation schemes are broadly divided into two categories—internal validation schemes and external validation schemes.

Internal validation can be defined as the estimation of the predictive accuracy of a model in the same study, whereas external validation is a validation done using truly independent data external to the study. Naturally, this type of validation depends on the modeling and the type of problem under study. Thus, external

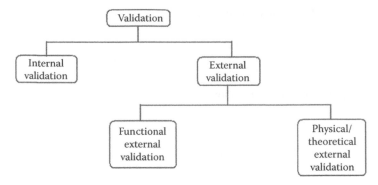

FIGURE 17.1   Types of validation schemes.

validation is more difficult than internal validation. Therefore, in the existing literature, more emphasis is on external validation techniques [2] than on internal validation techniques. Thus, this section mainly discusses the various external validation schemes.

External validation can be further categorized based on various criteria. The prime bases for the categorization of external validation are provided as follows:

1. Qualitative comparisons of dynamic response data
2. System identification techniques
3. Parameter sensitivity

In addition, there are categories based on the inverse models, the concepts from knowledge-based systems fields, and the role of expert opinions.

The three prime categories are discussed in Sections 17.4.1 through 17.4.3.

17.4.1   External Validation Based on Qualitative Comparisons of Dynamic Response Data

There are many methods under this category. Some of them are as follows:

1. Summation of the square of the difference between a system variable (y_i, i: time index) and the respective model variable (z_i). The following expression represents the method:

$$I = \sum_{i=1}^{n} (y_i - z_i)^2 \tag{17.16}$$

where:
   n is the total number of samples

2. Theil's inequality coefficient (TIC)—It can be used to get the difference between the system variable and the model variable in a better way. TIC is given by the following expression:

$$\text{TIC} = \frac{\sqrt{\sum_{i=1}^{n} (y_i - z_i)^2}}{\sqrt{\sum_{i=1}^{n} y_i^2} \sqrt{\sum_{i=1}^{n} z_i^2}} \tag{17.17}$$

3. The fitness functions used in genetic algorithms or simulated annealing.
4. Statistical methods
   a. Use of an autoregressive integrated moving average (ARIMA) or other stochastic models for time series data
   b. Step-wise regression procedures
   c. Use of multiple correlation coefficients
   d. Use of a hypothesis testing method
   e. Use of a spectral method
5. Model distortion approach

Distortion may be defined here as the deviation of the real system response from that of a model. If the distortion falls within a specified limit, then the model is said to be satisfactory.

17.4.2 EXTERNAL VALIDATION BASED ON SYSTEM IDENTIFICATION TECHNIQUES

The system identification process is considered to be the basis of this category because of the firm linkage between system identification and model validation. Identification can be categorized into structural identification and numerical identification.

A model is said to be structurally identifiable if a set of equations relating to a set of transfer function coefficients has a unique solution. If no unique solution is obtained, then the model is not structurally identifiable.

Numerical identifiability is determined by calculating the "parameter information matrix" M, which is given by

$$M = X^T X \tag{17.18}$$

where:
X is a parameter sensitivity matrix

The determinant of M is calculated and the numerical identifiability is determined either by this determinant or by a "condition number," which is defined as the ratio of the largest to the smallest eigenvalue of M. The confidence of the parameter estimates is large if the determinant is small or the condition number is large.

17.4.3 CATEGORY BASED ON PARAMETER SENSITIVITY

Validation by this category is not common nowadays, although the category may be effective in some cases. The method used here is the parameter sensitivity method that uses sensitivity functions by which the level of dependence between the response time history and the model parameters can be found.

17.5 APPLICATION AREAS OF CONTINUOUS SYSTEM SIMULATION

Models of continuous system simulation have been applied in physics, chemistry, aeronautics, bioinformatics, homeland security system, social systems, and so on. The areas of applications are increasing continuously as is evident from the existing literature.

Some examples of the application areas that have been dealt with up to now are as follows [3]:

1. Diffusion processes
2. Chemical plants
3. Digital computers
4. Analog systems
5. Autopilot studies
6. Flight systems
7. Various kinds of control loops for missiles
8. Aerodynamics
9. Power plant applications
10. Power distributions
11. Electrical machines
12. Dynamics of accidents of vehicles
13. Hydraulic systems
14. Spread of harmful substances

17.6 EVOLUTION OF CSSLs

The concept of continuous system simulation is a very traditional one, which is evident from the existing literature. However, because of the difficulties of the hardcore mathematics and the difficulty in implementing them, various software packages or languages have been developed since the infancy of the concept.

The first significant development of a language was accomplished in 1955. The name of the language was Selfridge and it was used to formulate block diagrams in digital computers. After that, a compiler named "Rose" was developed in 1959. Rose is basically a compiler with analog-oriented input language.

The significant change in the development of these CSSLs happened with the development of the MIDAS language that could be run on IBM 7090 series of computers. After that, COBLOC in 1963 and PACTOLUS in 1964 were developed. PACTOLUS is a predecessor of the Continuous Simulation Modeling Program (CSMP) language that was developed later in 1967. PACTOLUS could be run on

IBM 1620 computers. Next, DSL/90 was developed in 1965. It could translate its code into FORTRAN language formats. It could be run on IBM 7090 series of computers.

The real CSSL was first developed in 1967 with the development of the CSMP language. Later, the modified version of CSMP was developed in 1971. Another language called MIMIC was developed in 1968. This language could be run on Control Data Corporation (CDC) computers.

Mitchell and Gauthier Associates developed a continuous system simulation package in 1975, known as Advanced Continuous Simulation Language (ACSL), which was flexible enough to run on several models of computers. A modified version of ACSL, known as ECSSL, was developed for hybrid computers in 1978. In 1981, a block-oriented simulation language, known as EASYS, was developed. Later in 1980, many remarkable packages were developed. The names of these packages are ISIS/ISIM (1982), DSL (1984), and SYSMOD (1985).

However, with the advent of computer hardware and various other related technologies, new concepts have emerged that gave rise to a new generation of simulation packages for continuous system simulations. Some of the significant ones are described in Section 17.9. But before looking at some of these languages, it is better to understand their main features first. Thus, Section 17.7 lists the main features of the various CSSLs in brief.

17.7 FEATURES OF CSSLs

CSSL is applicable where one or more variables are continuous in nature. If some of the variables are discrete and others are continuous in nature, then a mixed approach combining both the discrete and continuous simulation concepts is applied. However, a true CSSL should contain the following basic features [3], which are basically the general features that are applicable to any generalized system developed:

1. Its syntax should be machine independent so that it can be used on different machines.
2. The provisions should be provided in language so that most of the parameters can be changed by the user of the system.
3. The modeling efforts should not be restricted by software capabilities.
4. The user should be able to run the simulation model in either an interactive mode or a batch mode. This feature is particularly difficult to implement from a development point of view, since these two provisions demand the development of a language from two completely different approaches.
5. The language should be easy to use and learn. It should be developed in such a way that novice users as well as experienced users can use it effectively.
6. These languages are developed for various scientific and research purposes, and those using these languages may not have much programming skill. Therefore, model development with these languages will be effective, not demanding much programming effort.

However, the features also depend on the type of CSSL used. Section 17.8 provides a brief overview of the types of CSSLs as is evident from the existing literature.

17.8 TYPES OF CSSLs

A continuous system simulation is useful when the behavior of the system depends more on an aggregate flow of events rather than on the occurrence of individual events. The existing literature shows a total of over 50 CSSLs. These languages can be categorized into group I and group II languages [4].

Group I languages are basically general CSSLs designed to represent dynamic models defined by a set of differential equations. Examples of such languages include MIMIC, DYNAMO, PROSE, CSMP/CSMP III, CSSL-III, and SL-I.

Group II languages are basically designed to represent models as defined by partial differential equations. Examples of such languages include SALEM, PDEL, LEANS, DSS, PDELAN, and FORSIM.

To formulate the directions to develop these languages, a committee was formed in the early 1960s, within the simulation councils. The task of this committee was to formulate and provide directions for language development of CSSLs. The features of the simulation languages for continuous system simulation were developed by this committee.

Section 17.9 briefly describes some of the most significant languages to provide a glimpse into such types of languages.

17.9 INTRODUCTION TO SOME CSSLs

In this section, a brief introduction to some of the significant CSSLs is provided. The main features are provided for each of these languages. There are separate sets of languages to deal with the ordinary differential equations and the partial differential equations. Therefore, Sections 17.9.1 and 17.9.2 provide a brief description of the languages for both ordinary differential equations and partial differential equations.

17.9.1 Languages for Ordinary Differential Equations

There are several languages in this category, some of which are as follows:

1. MIMIC
2. DYNAMO
3. CSMP-III
4. SL-1
5. PROSE
6. SLANG

The features of each language are described in Sections 17.9.1.1 through 17.9.1.6.

17.9.1.1 MIMIC

The overall features of MIMIC are as follows:

1. It is a nonprocedural language.
2. It contains FORTRAN IV modules.
3. Some of the simulation functions contained in MIMIC are LIMITER, TRACK, STORE, ZERO ORDER, HOLD, DEAD SPACE, TIME DELAY, and RANDOM NUMBER.
4. The Runge–Kutta method is used to perform integration.
5. MIMIC programs take longer to code and debug due to very rigid programming requirements and low diagnostic capability.
6. It could be run on CDC 6000 series of computers.

Its basic operations are delineated as follows:

1. The MIMIC source code is compiled into an intermediate format.
2. The intermediate program is sorted in an executable order.
3. The machine code is generated from the intermediate format.
4. The generated machine code is executed.

17.9.1.2 DYNAMO

The basic features of DYNAMO are delineated as follows:

1. It was originally developed in 1959 by the industrial dynamics group at the Massachusetts Institute of Technology (MIT) and later modified in 1965.
2. It can be run on IBM 360/370 computers.
3. It is a nonprocedural language.
4. A special unique subscription notation is used to keep track of time.
5. A macro is provided for simulation operators.
6. It has a good user-oriented diagnostic system.
7. It is easy to use and learn.
8. It may give misleading results due to "numerical analysis type problems."

17.9.1.3 CSMP-III

The main features of CSMP-III are described as follows:

1. It is an expression-based language.
2. It is a nonprocedural language.
3. It is a modified and improved version of CSMP/360.
4. It converts the source code to a FORTRAN-based intermediate program.
5. Both the system-defined and user-defined macros are provided in CSMP-III.
6. It uses seven to nine different integration algorithms.
7. It is easy to use and learn.
8. It has a better diagnostic and debugging system.
9. It can be run on CDC computers.

17.9.1.4   SL-1

The main features of SL-1 are provided as follows:

1. It is a nonprocedural language.
2. It was developed in 1970 by Xerox Data Systems.
3. It translates the source code to a FORTRAN language program that is translated to an assembly language (SYMBOL), which is finally translated to a machine code.
4. Both the FORTRAN and SYMBOL language constructs can be used in SL-1.
5. Its syntax is similar to that of CSMP.
6. It can be run on Xerox $\sum 5$ and $\sum 7$ computers using Xerox FORTRAN.
7. It is not easy to learn.

17.9.1.5   PROSE

The main features of PROSE are as follows:

1. It is a procedural simulation language.
2. It is a mathematics-based language.
3. All levels of users can use this language.
4. It is easy to learn, it has more diagnostic capability, and debugging is easier.
5. It can be applied to iterative techniques.
6. It can perform nonlinear optimizations and numerical integrations.
7. It can deal with both discrete and continuous simulations.
8. It is applicable to a wide range of simulation problems.

17.9.1.6   SLANG

The main features of SLANG are as follows:

1. It is a predecessor of PROSE.
2. It is a procedural simulation language.
3. It is a mathematics-based language.
4. All levels of users can use this language.
5. It is easy to learn, it has more diagnostic capability, and debugging is easier.
6. It can be applied to iterative techniques.
7. It can perform nonlinear optimizations and numerical integrations.
8. It can deal with both discrete and continuous simulations.

17.9.2   LANGUAGES FOR PARTIAL DIFFERENTIAL EQUATIONS

The main differences between these languages lie in the following:

1. The capability of the language in handling the classes and subclasses of partial differential equations
2. Flexibility in selecting an algorithm for a task to be accomplished
3. The type of language, for example, procedural or nonprocedural
4. The internal language used by these packages

Some of the significant languages dealing with partial differential equations are described briefly in Sections 17.9.2.1 through 17.9.2.5.

17.9.2.1 SALEM

The main features of SALEM are listed as follows:

1. It is a nonprocedural language.
2. It was developed at Lehigh University, Bethlehem, Pennsylvania.
3. It is no longer in use.
4. It consists of several FORTRAN subroutines and a translator.
5. It is applicable to one-dimensional (1D) parabolic, elliptic, and hyperbolic equations, and 2D parabolic and elliptic equations.
6. A tridiagonal algorithm is applied for 1D problems and an alternating direction technique is applied for 2D problems.
7. It has an effective diagnostic capability.

17.9.2.2 PDEL

The main features of PDEL language are given as follows:

1. It is a nonprocedural language.
2. It was developed in 1967 at the University of California.
3. It is a modular language.
4. It is a PL/I-based language.
5. It is applicable to 1D, 2D, or three-dimensional (3D) linear and nonlinear elliptic and parabolic partial differential equations, and 1D hyperbolic equations.
6. A tridiagonal algorithm is applied for 1D problems and an alternating direction technique is applied for 2D and 3D problems.

17.9.2.3 LEANS

The main features of LEANS are given as follows:

1. It is a nonprocedural language.
2. It was developed and is maintained by Lehigh University.
3. It shows greater capability in handling the classes and subclasses of partial differential equations compared to SALEM and PDEL.
4. It is more flexible than SALEM and PDEL in terms of selecting an algorithm for performing a specific task.
5. It is less user friendly than SALEM and PDEL.
6. FORTRAN IV is used as the base language.
7. It is applicable to 1D, 2D, and 3D elliptic, parabolic, and hyperbolic equations, and Cartesian, cylindrical, and spherical coordinates.
8. It uses differential–difference algorithms.

17.9.2.4 DSS

The main features of DSS are delineated as follows:

1. It was developed at Lehigh University.
2. It uses FORTRAN as base language.
3. It is applicable only to elliptic and parabolic partial differential equations.
4. A tridiagonal algorithm is applied for 1D problems and the Peaceman–Rachford algorithm is applied for 2D problems.
5. It has an effective diagnostic capability for input data errors and convergence failures.
6. It is capable of handling five or two simultaneous partial differential equations for 1D and 2D problems, respectively.

17.9.2.5 PDELAN

The main features of PDELAN are depicted as follows:

1. It was developed at the National Center for Atmospheric Research in Boulder, Colorado.
2. It is a procedural language.
3. It requires the user to have knowledge of finite difference algorithms.
4. It is applicable to simultaneous, nonlinear, and parabolic partial differential equations.
5. Running time for a program written in PDELAN is high compared to that written in FORTRAN.

Besides the above languages, there are also other languages such as FORSIM [4], XDS SIGMA 7, DARE M [5], GODYS-PC [6], AHCSSL [7], and EARLY DESIRE [8].

17.10 CONCLUSION

The various aspects of continuous system simulation have been discussed. The aspects that have been covered in this chapter include the various approaches applied to model a continuous system simulation problem. The various integration techniques and external validation schemes have also been discussed. Some of the significant application fields have been mentioned. Then, the various aspects of related languages observed in the existing literature have been presented. The features and categories of CSSLs have been presented in brief. Some of the significant languages for both ordinary differential equations and partial differential equations have also been introduced in this chapter.

REFERENCES

1. Cellie, F. E. and Kofman, E. (2006). *Continuous System Simulation*. New York: Springer.
2. Murray-Smith, D. J. (1998). Methods for the external validation of continuous system simulation models: A review. *Mathematical and Computer Modelling of Dynamical Systems: Methods, Tools and Applications in Engineering and Related Sciences* 4(1): 5–31.
3. Bausch-Gall, I. (1987). Continuous system simulation languages. *Vehicle System Dynamics: International Journal of Vehicle Mechanics and Mobility* 16(suppl. 1): 347–366.

4. Nilsen, R. N. and Karplus, W. J. (1974). Continuous-system simulation languages: A state-of-the-art survey. *Mathematics and Computers in Simulation* 16: 17–25.
5. Korn, G. A. (1974). Recent computer system developments and continuous system simulation. *Mathematics and Computers in Simulation* 16: 2–11.
6. Kuras, J., Lembas, J., and Skomorowski, M. (1997). Godys-PC: An interactive continuous system simulation language. *Simulation Practice and Theory*, 20: 32–33.
7. Mawson, J. B. (1975). A continuous system simulation language for an advanced hybrid computing system (AHCSSL). *International Symposium on Simulation Languages for Dynamic Systems*, September 8–10, London: WorldCat, pp. 283–293.
8. Korn, G. (1982). Early desire. *Mathematics and Computers in Simulation*, 24(1): 30–36.

18 Introduction to Simulation Optimization

18.1 INTRODUCTION

The definition of simulation optimization is difficult to present because of the various uses, aspects, and treatments for the concept. Research studies in the existing literature have not distinguished among the various aspects of such optimization methods, although the various studies in the literature reveal the aspects. One of these aspects is the application of the various optimization techniques on the various simulation parameters. Another aspect is to use simulation to optimize a system.

However, there is a significant number of studies on simulation optimization observed in the existing literature. An August 2013 search for research papers by the keyword "simulation optimization" showed a total of 2,617 research studies in the ScienceDirect database; 1,335 journal papers and book chapters in the Springer database; 797 research papers in the Taylor & Francis database; 367 research papers in the IEEE database; 229 research papers in the JSTOR database; and a total of 21,200 results in Google Scholar. These search results clearly indicate the increasing interest in the area of simulation optimization among researchers all over the world. The main reason behind such enthusiasm seems to be the challenges and difficulties faced by the researchers of this particular field of study.

The main difficulty lies in configuring the parameters of a simulation study to optimize them. The features of various software available in the existing market also pose a big problem. Since the source codes of commercial software are very difficult to obtain, embedding optimization techniques into a simulation package poses a big problem for such research studies. However, recent progress in the literature is observed. There is an endeavor to propose tools for bridging the gap between a simulation package and optimization algorithms. Moreover, many of the commercial packages include optimization modules in their simulation software nowadays. However, the effort toward resolving this difficulty has not been significant and therefore the difficulty still persists.

However, this chapter discusses the various aspects of the concept of simulation optimization. And in Chapter 19, some of the optimization tools used in case simulation are presented. The sections of this chapter are arranged as follows: Section 18.2 discusses the various aspects of optimization for simulation. Section 18.3 discusses the major issues observed while conducting research studies in this area and some of the major advantages of simulation optimization. Section 18.4 introduces some major commercial packages on simulation optimization. Section 18.5 outlines the application areas of simulation optimization. Section 18.6 provides the conclusion for this chapter.

18.2 ASPECTS OF OPTIMIZATION FOR SIMULATION

The main aspects discussed in Section 18.1 are the approaches used in applying simulation and optimization techniques. Some of the research studies have applied optimization tools to optimize the output of a simulation experiment [1]. In this case, the main difficulty lies in embedding the optimization tools in the simulation package to be used. This approach is depicted in Figure 18.1.

Various optimization tools can be used for optimization purposes in this case. But since the source code of commercial simulation packages is not easy to obtain, embedding an optimization tool into a commercial simulation package is difficult. In addition, the simulation parameters are also required to be configured properly in such cases. If the optimization tool is used to optimize some input parameters along with the output of simulation, then the problem becomes more complex.

The second aspect of simulation optimization is the application of a simulation package to optimize a system (Figure 18.2). Thus, in this case, the endeavor is to maximize or minimize the variables of a formulated optimization problem with the help of simulation [1].

The main difficulty in solving the second type of problem lies in the simulation experimentation itself. While developing a system or flowchart for a simulation, one mainly considers a particular scenario or situation. Thus, a simulation is usually generalized to be applicable to all types of problems on the topic. For example, let us consider a parallel machine scheduling problem for a manufacturing shop floor. In this case, a particular machine arrangement is generally considered before conducting a simulation study. Although the arrangement can be changed, separate sets of simulation studies will have to be conducted for the various machine setups that are to be considered. This means that a particular simulation is not applicable to all situations. In other words, the simulation is not generalized. Therefore, the generalized simulation poses a challenge in the second type of simulation optimization problem.

However, to understand the concept of simulation optimization, Section 18.3 raises various issues that must be understood thoroughly.

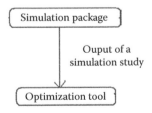

FIGURE 18.1   Optimization of output of simulation study.

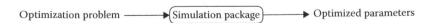

FIGURE 18.2   Use of simulation for optimization purpose.

18.3   MAJOR ISSUES AND ADVANTAGES
OF SIMULATION OPTIMIZATION

Simulation optimization is a challenging research topic from both the aspects discussed in Section 18.2. To facilitate research studies in this field, the following issues need the intense attention of researchers:

1. It is necessary to combine a simulation package with an optimization tool or, in other words, embed an optimization tool in a simulation package. Although recent versions of most simulation packages have optimization modules, it is not possible to combine just any optimization tool with any simulation package, and therefore it is necessary to develop a common platform.
2. Simulation packages are generally more expensive to run on a computer than analytical functions. This issue poses another challenge for research studies in this field.
3. If the objective functions and/or constraints are stochastic or fuzzy in nature, then simulation becomes more difficult. Thus, further investigations into these issues are necessary.

However, in spite of having serious difficulties and challenges, the use of simulation for optimization has the following advantages:

1. The handling of a complex system becomes easier with the help of simulation.
2. Numerous output analysis techniques are available that can be used to control the variance of output for a stochastic or fuzzy system.
3. The application of simulation techniques makes structural optimization easier.

18.4   COMMERCIAL PACKAGES FOR
SIMULATION OPTIMIZATION

Nowadays, most simulation packages include optimization modules to facilitate simulation optimization. Reviews of such commercial packages are found in the existing literature [2,3]. Some of these packages are as follows:

1. *AutoMod*. The features of AutoMod software are as follows:
   a. The product suite includes AutoMod for simulation, AutoStat for analysis, and AutoView for 3D animation purposes.
   b. This software was developed by AutoSimulations Inc., California.
   c. The main application areas of this software are manufacturing and material handling systems.
   d. The model consists of a system that can either be a process system (for manufacturing applications) or a movement system (for material handling applications).

   e.   A model developed using AutoMod can be reused as an object in other
        models.
   f.   The optimization strategies used in AutoMod are evolutionary strate-
        gies and genetic algorithms.
   g.   The user can build scenarios, conduct experimentation, and perform
        analyses.
2. *OptQuest*. The features of OptQuest optimizer are as follows:
   a.   It is a field installation software engine.
   b.   It can be used in Arena and Crystal Ball.
   c.   It was developed by Optimization Technologies, Inc.
   d.   It uses search techniques like scatter search, tabu search, and neural
        networks.
   e.   It can solve various optimization problems that cannot be solved by the
        standard Excel solver.
   f.   It can handle problems with large numbers of variables and constraints.
3. *OPTIMIZ*. The basic features of OPTIMIZ are as follows:
   a.   It can simulate the performance of an entire plant.
   b.   It was developed by Visual Thinking International Ltd.
   c.   It allows visualization of input data and interpolation curves.
   d.   A model developed using OPTIMIZ can be used in other models devel-
        oped using OPTIMIZ.
   e.   The optimization tool used by OPTIMIZ is a neural network.

In addition, there are several other packages such as SimRunner (developed by
PROMODEL Corporation Orem, Utah) and Optimizer (developed by Lanner Group
Inc., UK).

18.5   APPLICATION AREAS OF SIMULATION OPTIMIZATION

Simulation optimization has been applied to various fields of study as evident from
the existing literature.

The literature shows various types of applications of simulation optimization.
Simulation optimization has been used in various concepts such as multicriteria decision
analysis [4]. Other application areas include decision support systems [5], scheduling
problems [6], inventory systems [7], groundwater management [8], manufacturing
[9], relational database analysis [10], supply chain management [11], investment evaluation [12], biotechnological processes [13], and physics [14]. Thus, various fields of
study have benefited from the application of simulation optimization.

18.6   CONCLUSION

In this chapter, the various aspects of simulation optimization have been discussed.
Various applications, advantages, issues, and commercial software packages have
been introduced in this chapter. Simulation optimization is a very challenging field
of study. Although numerous research studies have been published that have mainly
made use of various optimization algorithms and metaheuristics, optimization aided

by simulation is still a very challenging area of research and very few papers have been published on it. Therefore, there is ample scope of doing further research in this area.

REFERENCES

1. Nelson, B. L. (2010). Optimization via simulation over discrete decision variables. *INFORMS TutORial in Operations Research*, pp. 193–207, doi:10.1287/educ.1100.0069.
2. April, J., Glover, F., Kelly, J. P., and Laguna, M. (2003). Practical introduction to simulation optimization. *Proceedings of the 2003 Winter Simulation Conference*, December 7–10, New Orleans, LA. New York: ACM, pp. 71–78.
3. Fu, M. C. (2002). Optimization for simulation: Theory vs. practice. *INFORMS Journal on Computing* 14(3): 192–215.
4. Kuriger, G. W. and Grant, F. H. (2011). A Lexicographic Nelder-Mead simulation optimization method to solve multi-criteria problems. *Computers & Industrial Engineering* 60(4): 555–565.
5. Li, Y. F., Ng, S. H., Xie, M., and Goh, T. N. (2010). A systematic comparison of metamodeling techniques for simulation optimization in decision support systems. *Applied Soft Computing* 10(4): 1257–1273.
6. Yang, T. (2009). An evolutionary simulation-optimization approach in solving parallel-machine scheduling problems: A case study. *Computers & Industrial Engineering* 56(3): 1126–1136.
7. Hochmuth, C. A. and Köchel, P. (2012). How to order and transship in multi-location inventory systems: The simulation optimization approach. *International Journal of Production Economics* 140(2): 646–654.
8. Gaur, S., Chahar, B. R., and Graillot, D. (2011). Analytic element method and particle swarm optimization based simulation-optimization model for groundwater management. *Journal of Hydrology* 402(3/4): 217–227.
9. Dengiz, B., Bektas, T., and Ultanir, A. E. (2006). Simulation optimization based DSS application: A diamond tool production line in industry. *Simulation Modelling Practice and Theory* 14(3): 296–312.
10. Willis, K. O. and Jones, D. F. (2008). Multi-objective simulation optimization through search heuristics and relational database analysis. *Decision Support Systems* 46(1): 277–286.
11. Chen, Y., Mochus, L., Orcan, S., and Reklaitis, G. V. (2012). Simulation-optimization approach to clinical trial supply chain management with demand scenario forecast. *Computers & Chemical Engineering* 40(11): 82–96.
12. Ramasesh, R. V. and Jayakumar, M. D. (1997). Inclusion of flexibility benefits in discounted cash flow analyses for investment evaluation: A simulation optimization model. *European Journal of Operational Research* 102(1): 124–141.
13. Brunet, R., Guillén-Gosálbez, G., Pérez-Correa, J. R., Caballero, J. A., and Jiménez, L. (2012). Hybrid simulation-optimization based approach for the optimal design of single-product biotechnological processes. *Computers & Chemical Engineering* 37: 125–135.
14. Li, R., Huang, W., Du, Y., Shi, J., and Tang, C. (2011). Simulation optimization of single-shot continuously time-resolved MeV ultra-fast electron diffraction. *Nuclear Instruments and Methods in Physics Research Section A: Accelerators, Spectrometers, Detectors and Associated Equipment* 637(1): S15–S19.

19 Algorithms for Simulation Optimization

19.1 INTRODUCTION

In Chapter 18, an introduction to the concept of simulation optimization was presented. To implement the concept of simulation optimization in various research studies, the authors in the existing literature have used various techniques. Although the optimization techniques for simulation optimization are not limited to the optimization techniques described in this chapter, a brief introduction to the various techniques used in the existing literature is necessary. The optimization techniques that have been applied to simulation optimization highlight the gap in research in this area in terms of their applications.

This chapter briefly presents the various optimization techniques discussed in the existing literature. The remaining sections of this chapter are arranged as follows: Section 19.2 provides a glimpse of the major techniques that have been applied to the various fields of study. Section 19.3 provides an introduction to some other techniques. Section 19.4 provides the conclusion for this chapter.

19.2 MAJOR TECHNIQUES

Most of the published research studies have mentioned mainly four optimization techniques that have been used in simulation optimization. These techniques are as follows:

1. Gradient-based search techniques
2. Stochastic approximation
3. Response surface methodology (RSM)
4. Heuristic methods

In addition, Fu [1] has classified the major techniques into two types: (1) discrete parameter cases and (2) continuous parameter cases. The techniques included in the discrete parameter cases are as follows:

1. Multiple comparison procedure
2. Ranking-and-selection procedure

The techniques included in the continuous parameter case are gradient-based methods that can further include the following techniques:

1. Perturbation analysis
2. Likelihood ratio (LR) method
3. Frequency domain experimentation

A set of techniques has also been provided by Andradóttir [2]. These techniques are as follows:

1. Gradient-based methods
2. Simulated annealing (SA)
3. Genetic algorithm (GA)
4. Evolutionary algorithms
5. Artificial neural network
6. Scatter search
7. Tabu search (TS)
8. Random search
9. Sample path optimization
10. RSM
11. Ranking and selection
12. Cross-entropy method

A picturesque classification in a hierarchical form has been presented by Carson [3]. This chapter focuses on the commonly used techniques that are listed in Figure 19.1.

The four major optimization techniques are described briefly in Sections 19.2.1 through 19.2.4.

19.2.1 Gradient-Based Search Techniques

Gradient-based search methods can be applied to nonlinear problems. The various methods of estimating gradients are used in simulation optimization to decide on the shape of the objective function. The successful application of this method depends on its efficiency and reliability. Reliability is important since significant errors in gradient estimation techniques may lead to the wrong direction and thus the wrong result. In addition, the response of a simulation study is stochastic in nature. The simulation experiments are very expensive to run in computers. Therefore, if the method is efficient, then it will result in less effort, which in turn will result in less expense. Hence, efficiency is important. However, the main techniques in this category that are used for simulation optimization are as follows:

1. Finite difference estimation
2. LR estimators
3. Perturbation analysis
4. Frequency domain experiments

Each of these methods is described briefly in Sections 19.2.1.1 through 19.2.1.4.

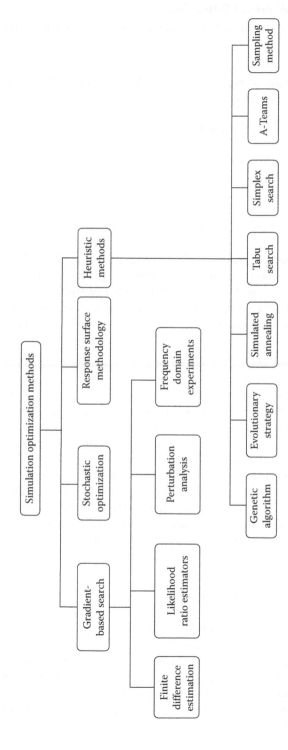

FIGURE 19.1 Various optimization techniques used in simulation optimization.

19.2.1.1 Finite Difference Estimation

The finite difference of a variable x is a mathematical expression given by

$$f(x+b) - f(x+a)$$

The difference quotient is obtained if this expression is divided by $(b - a)$. There are three types of finite differences:

$$\text{Forward difference} = \frac{f(x+h) - f(x)}{h} \tag{19.1}$$

$$\text{Backward difference} = \frac{f(x) - f(x-h)}{h} \tag{19.2}$$

$$\text{Centered difference} = \frac{f(x+h) - f(x-h)}{2h} \tag{19.3}$$

The finite difference estimation method is the crudest method of estimating the gradient. This estimation is used in calculating the gradient at a point in the solution space, which is found by a predecided number of simulation runs. Here, the response of a simulation run is stochastic in nature. The disadvantage of this method is that the estimation of an objective function by this method may be noisy and thus wrong.

19.2.1.2 LR Estimators

LR is commonly known to be a statistical test. In general, an LR test provides the means for comparing the likelihood of the data under one hypothesis (usually called the *alternate* hypothesis) with the likelihood of the data under another, more restricted hypothesis (usually called the *null* hypothesis, for the experimenter tries to *nullify* this hypothesis to provide support for the former).

However, with gradient-based search methods, the LR estimation is the gradient of the expected value of output variable with respect to an input variable and is given by the expected value of a function of input parameters or the expected value of a function of simulation parameters, such as simulation run length. This method is more appropriate for transient and regenerative simulation optimization problems.

19.2.1.3 Perturbation Analysis

In infinitesimal perturbation analysis (IPA), all partial gradients of an objective function are estimated from a single simulation run. The idea is that in a system, if an input variable is perturbed by an infinitesimal amount, the sensitivity of the output variable to the parameter can be estimated by tracing its pattern of propagation. This will be a function of the fraction of propagations that die before having a significant effect on the response of interest. IPA assumes that an infinitesimal perturbation in an input variable does not affect the sequence of events but only makes its number of occurrences slide smoothly. The fact that all derivatives can be derived from a single simulation run represents a significant advantage in terms of computational efficiency. However, estimators derived using IPA are often biased and inconsistent. The IPA gradient estimator is more efficient than the LR method. Other IPA methods include smoothed perturbation analysis (SPA) and IPA variants.

19.2.1.4 Frequency Domain Experiments

A frequency domain experiment is one in which the selected input parameters are oscillated sinusoidally at different frequencies during one long simulation run. The output variable values are subjected to spectral (Fourier) analysis, that is, regressed against sinusoids at the input driving frequencies. If the output variable is sensitive to an input parameter, the sinusoidal oscillation of that parameter should induce the corresponding (amplified) oscillations in the response.

Frequency domain experiments involve addressing the three following questions: How does one determine the unit of the experimental or oscillation index? How does one select the driving frequencies? And how does one set the oscillation amplitudes?

19.2.2 STOCHASTIC OPTIMIZATION

Stochastic optimization is the problem of finding a local optimum for an objective function whose values are not known analytically but can be estimated or measured. Classical stochastic optimization algorithms are iterative schemes based on gradient estimation. Proposed in the early 1950s, Robbins–Monro and Kiefer–Wolfowitz are the two most commonly used algorithms for unconstrained stochastic optimizations. These algorithms converge extremely slowly when the objective function is flat and often diverge when the objective function is steep. Additional difficulties include the absence of good stopping rules and handling constraints.

In general, stochastic optimization methods make use of random variables. The random variables may be present in the objective functions and/or constraint functions. The various stochastic optimization methods may include the following techniques:

1. Stochastic optimization proposed by Robbins and Monro [4]
2. Stochastic gradient descent
3. Simultaneous perturbation stochastic optimization
4. Scenario optimization
5. Finite difference stochastic optimization

19.2.3 RESPONSE SURFACE METHODOLOGY

RSM is a collection of statistical and mathematical techniques useful for developing, improving, and optimizing processes. It also has important applications in the design, development, and formulation of new products, as well as in the improvement of existing product designs.

The most extensive applications of RSM are in the industrial world, particularly in situations where several input variables have the potential to influence some performance measure or quality characteristic of the product or process. This performance measure or quality characteristic is called the *response*. It is typically measured on a continuous scale, although attribute responses, ranks, and sensory responses are not unusual. Most real-world applications of RSM involve more than one response. The input variables are sometimes called *independent variables*, and they are subject to the control of the engineer or scientist, at least for the purpose of a test or an experiment.

The practical application of RSM requires developing an approximating model for the true response surface. The underlying true response surface is typically driven by some unknown *physical mechanism*. The approximating model is based on the observed data from the process or system and is an *empirical model*. A multiple regression is a collection of statistical techniques useful for building the types of empirical models required in RSM.

19.2.4   HEURISTIC METHODS

Numerous heuristic methods exist in the existing literature. Some of them are briefly discussed in Sections 19.2.4.1 through 19.2.4.5.

19.2.4.1   Genetic Algorithm

GAs are a kind of metaheuristic method developed based on the natural process of genetic reproduction. GAs, proposed by John Holland and his team at the University of Michigan, Ann Arbor, and later developed by Goldberg, have the largest share in the existing literature in terms of the number of research publications. GAs have been used widely to solve multiobjective problems. Numerous algorithms in this direction are observed.

In GA terms, the DNA of a member of a population is represented as a string where each position may take on a finite set of values. Each position in the string represents a variable from the system of interest, for example, a string of five input switches on a black box device where each switch may take the value 1 (switch is on) or 0 (switch is off); the string 11100 indicates the first three switches to be on and the last two switches to be off. The fitness of a member of a population is determined by an objective function. Members of a population are subjected to operators to create offspring. Commonly used operators include selection, reproduction, crossover, and mutation (see the work of Goldberg [5] for further details). Several generations may have to be evaluated before significant improvement in the objective function is seen. GAs are noted for robustness in searching complex spaces and are best suited for combinatorial problems.

19.2.4.2   Evolutionary Strategy

Similar to GAs, evolutionary strategies (ESs) are algorithms that imitate the principles of natural evolution as a method for solving parameter optimization problems. Rechenberg is credited for introducing ESs during the 1960s at the Technical University of Berlin in Germany.

The first algorithm employed was a simple mutation–selection scheme called the two-membered ES or $(1 + 1) - ES$. This scheme consisted of one parent producing one offspring by adding standard normal random variates. The better of the parent and the offspring becomes the parent for the next generation. The termination criteria include the number of generations, the elapsed central processing unit (CPU) time, the absolute or relative progress per generation, and so on. The multimembered ES, or $(m + 1) - ES$, involves two parents, randomly selected from the current population of $m > 1$ parents, producing one offspring. Extensions of the $(m + 1) - ES$ scheme include $(m + 1) - ES$ and $(m,1) - ES$. In an $(m + 1) - ES$, m parents produce one offspring followed by the removal of one least fit individual (parents and offspring) to restore the population size to m. An $(m,1) - ES$ is comparable to $(m + 1) - ES$; however, only the offspring will undergo selection. This was also elaborated on other complex versions of ES, such as correlated mutations, along with experimental comparisons of ES with popular direct search methods.

19.2.4.3 Simulated Annealing

SA [6] is a metaheuristic method that simulates the process of annealing where a solid is heated and then cooled continuously. Thus, the temperature is raised up to a certain level (higher energy state), and then the temperature is lowered to a certain level (lower energy state). SA is actually an adaptation of the Metropolis–Hastings algorithm (MHA), which is a Monte Carlo method used to generate sample states of a thermodynamic system. MHA is used to find a sequence of random samples from a probability distribution for which direct sampling is difficult. The generated sequence may be used to approximate the distribution.

The implemented single-objective optimization algorithm also depends on the temperature value. The solution is modified based on the current temperature. The two other important parameters of the algorithm are the cooling schedule and the number of iterations.

19.2.4.4 Tabu Search

The TS approach is based on the idea of accepting the nearest neighboring solution that has low cost, thus making it a local search procedure. A total of three types of memories are considered. The short-term memory of already found solutions is set as "Tabu" so that those solutions are never revisited. The intermediate memory is used to store the intermediate or current near-optimal solutions, and the long-term memory is used to record the search space areas that have already been searched. Thus, diverse solutions can be obtained by the proper use of the long-term memory.

A problem faced by TS is that the search may converge to a small area in the search space. To resolve this problem, the tabu list is made to consist of the attribute of a single solution instead of the entire solution, which results in another problem. When a single attribute is made a tabu, then more than one solution may become a tabu, and some of these solutions, which have to be avoided, might be better solutions but might not be visited. To solve this problem, "aspiration criteria" are used. An aspiration criterion overrides the tabu state of a solution and includes the better solution that could otherwise be excluded. An aspiration criterion is commonly used to allow solutions that are better than the current best solutions.

19.2.4.5 Simplex Search

The search starts with points in a simplex consisting of $p + 1$ vertices (not all in the same plane) in the feasible region. It proceeds by continuously dropping the worst point in the simplex and adding a new point determined by the reflection of the worst point through the centroid of the remaining vertices. Disadvantages of this method include the assumption of convex feasible regions and the implementation of problems involving the handling of feasibility constraints.

19.2.4.6 A-Teams

A-Teams is a type of heuristic algorithm that denotes asynchronous teams. The basic concept is to combine various problem-solving strategies (hybrid strategies) so as to fulfill the basic purpose of solving a problem. The various problem-solving strategies interact among one another synergistically. For example, one can combine GA with Newton's method to solve a multiobjective problem. A-Teams is especially suited to solving complex multiobjective or multicriteria problems. A combination of several algorithms results in a fast and robust solution procedure for complex problems.

19.2.4.7 Sampling Method

This method is especially used to emphasize rare events for a sample. Naturally, the sampling method used for this purpose is importance sampling. The basic idea is to use different probability measures from different probability distributions for the system as a whole. The probability of the sample containing rare events is increased so as to have a focus on that sample. Each of the probability measures for each sample path is multiplied by a correction factor to obtain an unbiased estimate. In this way, rare events are monitored; an example of such a rare event is an asynchronous transfer mode (ATM) communication network failure.

19.3 SOME OTHER TECHNIQUES

There are numerous other techniques for simulation optimization, as evident from the existing literature. Some of these techniques are outlined in Sections 19.3.1 and 19.3.2.

19.3.1 DIRECT Optimization Algorithm

The main features of the DIRECT optimization algorithm are as follows:

1. It is based on the Lipschitzian optimization algorithm.
2. It has been proposed to solve global optimization problems with bound constraints and real-valued objective functions.
3. It converges to a global optimum value.
4. It is basically a sampling algorithm.

5. Its full form is a dividing rectangle, which indicates that the algorithm divides the solution space into rectangles and proceeds toward the optima through these rectangles.
6. It starts its action by considering the problem domain as a unit hypercube.

19.3.2   UOBYQA OPTIMIZATION ALGORITHM

The main features of the UOBYQA algorithm are as follows:

1. Its full form is unconstrained optimization by quadratic approximation.
2. It uses interpolation to form quadratic functions for the objective function.
3. It is a derivative-free method.
4. It uses its iteration for minimizing the quadratic model and generates a new vector of variables.

19.3.3   SPLASH

The main features of Splash are as follows:

1. Splash stands for Smarter Planet Platform for Analysis and Simulation of Health. Splash 2 is a complete application suite that contains eight applications and four kernels, designed for various scientific, engineering, and graphics-based computing.
2. It is an application for modeling and simulating complex systems.
3. It is basically a prototype system.
4. It combines various complex heterogeneous simulation systems to develop a composite simulation model.
5. It uses various simulation and data integration technologies to integrate various heterogeneous systems.
6. There are various components of Splash. The experiment management component of Splash uses Rinott's procedure to select the best system under consideration.
7. Splash 2 has removed some of the codes of the original Splash.
8. The eight applications of Splash 2 are as follows:
   a. Barnes, which is used to simulate the interaction of a system of bodies
   b. The fast multipole method (FMM), which is used to simulate the interaction of systems of bodies over a number of time steps
   c. Ocean, which is used to simulate ocean movements with eddy and boundary currents
   d. Radiosity, which is used to simulate the distribution of lights in a scene
   e. Raytrace, which is used in ray tracing
   f. Volrend, which is used to simulate ray casting techniques
   g. Water-Nsquared, which is used to simulate forces and potentials among water molecules
   h. Water-spatial, which is used to simulate forces and potentials among water molecules using more efficient algorithms

9. The four kernels of Splash 2 are as follows:
   a. Cholesky, which is used for matrix calculations
   b. Fast Fourier transform (FFT), which is used for simulation of FFT algorithms
   c. LU, which is used for simplifying a dense matrix
   d. Radix, which is used for radix sort procedures

19.4 CONCLUSION

Basically, in this chapter, the algorithms used by various authors in their research studies in the existing literature have been discussed. The benchmark algorithms in this area have been outlined as a whole. The various algorithms mentioned here include gradient-based search methods, stochastic optimization methods, RSM, and various heuristic methods. A brief introduction to each of these methods has been presented in this chapter. Although there are several other methods, as evident from the existing literature, these algorithms have mostly been applied in the existing literature.

REFERENCES

1. Fu, M. C. (1994). Optimization via simulation: A review. *Annals of Operations Research* 53: 199–248.
2. Andradóttir, S. (1998). A review of simulation optimization techniques. *Proceedings of the 1998 Winter Simulation Conference*, December 13–16, Orlando, FL. New York: ACM, pp. 151–158.
3. Carson, Y. (1997). Simulation optimization: Methods and applications. *Proceedings of the 1997 Winter Simulation Conference*, December 7–10, Atlanta, GA. New York: ACM, pp. 118–126.
4. Robbins, H. and Monro, S. (1951). A stochastic approximation method. *The Annals of Mathematical Statistics* 22(3): 400–407.
5. Goldberg, D. E. (1989). *Genetic Algorithms in Search, Optimization & Machine Learning.* 5th Indian Reprint. New Delhi: Pearson Education.
6. Kirkpatrick, S., Gelatt, C. D., and Vecchi, M. P. (1983). Optimization by simulated annealing. *Science* 220(4598): 671–680.

20 Simulation with System Dynamics

20.1 INTRODUCTION

System dynamics is a field of study that focuses on the behavior of a system. A system may be identified as a collection of interrelated parts working together to achieve a common goal. For example, a family is a system that consists of the members of the family and the related components to keep the family members together.

As a system becomes more and more complex, the general description of the system or the mathematical theories that may be applicable to the system cannot represent the fragmented knowledge of the system, especially while all the fragments are interrelated.

System dynamics endeavors to organize the complex structure of such systems. Such a structure should clearly indicate the contradictions as well as the solutions to the problems inherent in the system.

Systems can be open or closed. A closed system is also known as a feedback system since it includes the feedback. Figure 20.1 shows an open system in which no feedback is shown. In an open system, the system does not learn from the output. In other words, the system is unable to rectify the flaws inherent in the system or to modify the existing system based on the output.

Figure 20.2 shows a closed system with feedback from output to input. Feedback enables learning for the system and improves the performance of the system. Therefore, the system can learn from the resulting output and accordingly a modification of the system will occur.

However, before explaining the aspects of simulation with system dynamics, it is necessary to clarify some related concepts. Therefore, Section 20.2 explains some of the basic concepts related to system dynamics.

20.2 IMPORTANT CONCEPTS RELATED TO SYSTEM DYNAMICS

To understand the broad concept of system dynamics, some important concepts need clarification. Thus, some of the important concepts of system dynamics are briefly presented in Sections 20.2.1 through 20.2.4.

20.2.1 FEEDBACK LOOP

A feedback loop is a kind of loop based on the feedback of several components of a system. Each of these components provides feedback to another component that in turn provides feedback to some other component in the system, which ultimately

FIGURE 20.1   Open system.

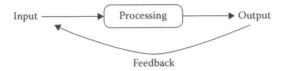

FIGURE 20.2   Closed system.

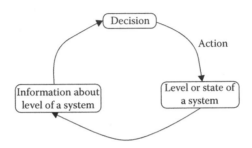

FIGURE 20.3   Feedback loop.

provides feedback to the initial component, thus giving rise to a loop. Figure 20.3 shows such a feedback loop [1].

Figure 20.3 shows a simple general feedback loop without any complexity, which indicates that the level or state of a system generates information about the system that in turn leads to decision making. As a result of decision making, action is triggered, which again changes the level or state of the system. This loop continues until the system stabilizes.

In a real-world system, there can be numerous feedback loops that can make the system more complex, for example, an inventory ordering loop. However, the difference between the true system level and the information level drives the decision making in a system. That is why the information level is sometimes called the apparent level.

20.2.2   Positive and Negative Feedback

There are two types of feedback—positive feedback and negative feedback. Positive feedback forces growth in a system, whereas negative feedback indicates the failure of a system. Negative feedback helps to find the causes of fluctuations, discrepancies, and instability in a system. Figure 20.4 shows an example of negative feedback.

The figure shows that an increase in carbon dioxide in the atmosphere results in an increase in temperature. As a result, water evaporation accelerates in the atmosphere, which leads to cloud formation. Rainfall occurs as a result of cloud formation, which in turn decreases atmospheric temperature and results in equilibrium.

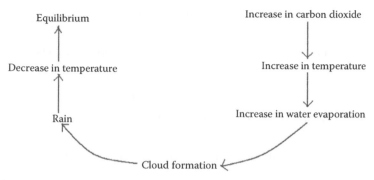

FIGURE 20.4 Example of negative feedback.

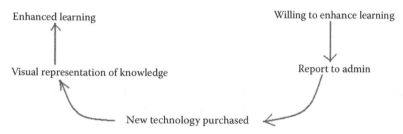

FIGURE 20.5 Example of positive feedback.

Here, the failure of the system indicates the temperature increase, which is the instability resulting because of increased carbon dioxide. This instability leads to the natural rectification in the atmosphere, which ultimately results in equilibrium. Forrester [1] gave a simple example of negative feedback. For example, a watch does not show the correct time. As a result of the failure of the clock, the user rectifies the time following a correct clock's time. Figure 20.5 shows an example of positive feedback.

Positive feedback increases the growth of a system. Figure 20.5 shows that a willingness to enhance learning or knowledge facilitates the visual representation of knowledge, which ultimately resultes in further growth or enhancement of knowledge or learning.

20.2.3 First-Order Negative Feedback Loop

Since a negative feedback loop indicates instability in the system that leads to a rectification, the negative feedback loop needs more attention. A negative feedback loop can be classified into a first-order negative feedback loop and a second-order negative feedback loop.

In a first-order negative feedback loop, only one decision controls a level or state of the system. Figure 20.6 shows a first-order negative feedback loop. The figure shows the scenario of a department of a university facing a scarcity of properly qualified teachers. As a result of teachers leaving the organization, the class load per teacher increases, which results in sickness and more absences of the teachers,

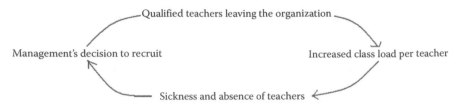

FIGURE 20.6   Example of first-order negative feedback loop.

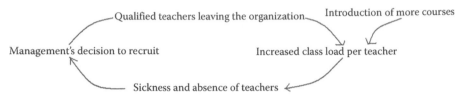

FIGURE 20.7   Example of second-order negative feedback loop.

which makes the management realize the need to recruit more teachers. Thus, here, only one decision, that is, that of the teachers leaving the organization, leads to the increase in class load. The word "first order" is used since only one decision is controlling the level (increased class load).

20.2.4   SECOND-ORDER NEGATIVE FEEDBACK LOOP

In the case of a second-order negative feedback loop, a total of two decisions control the level or state of the system. The word "second order" is used to represent the two decisions controlling the level. Figure 20.7 shows an example of a second-order negative feedback loop.

The figure shows a total of two decisions—qualified teachers leaving the organization and the introduction of more courses, which results in increased class loads per teacher. Since two decisions are controlling the level or state of the system, this is a second-order negative feedback loop.

Before delving further into the concept of system dynamics, we discuss in Section 20.3 a list of the steps required to perform a simulation study with the help of system dynamics.

20.3   STEPS IN MODELING WITH SYSTEM DYNAMICS

The steps of modeling with system dynamics provide important insights into the simulation study itself. To properly simulate the behaviors of a system, the steps depicted in the following will guide a user toward the goal of developing a proper simulation model in a systematic way. The steps are outlined as follows [2]:

Step 1: Define the problem
   The first question that can be asked while defining the problem under study
   is, What is happening over time in the system that we are about to deal

with? This means that this step endeavors to define the problem on hand. The question of why does the problem arises also searches for the causes of the problem encountered. In this stage, the key variables and key concepts of the system are also identified. Thus, modelers try to search for answers to the following questions:

- What is the problem?
- Why has the problem occurred?
- What are the key variables?
- What are the key concepts?
- What is the time horizon over which the solution of the problem is to be considered?

Step 2: List the factors

This step asks the following questions:

- What are the driving factors of the system?
- Do the factors explain any trend?
- Is there any other factor that should be considered?
- What are the model boundaries?
- What are the endogenous and exogenous consequences of the feedback structure?
- What are the current concepts about the system behavior?

Answers to the the above questions are essential in formulating a dynamic hypothesis for the system. These answers help in mapping and developing the simulation model.

Step 3: Map the system structure

Generalized maps of the structure of the system can be drawn now based on the identified variables, concepts, perceived behaviors, and system boundaries. Mapping of this causal structure can be accomplished by various system dynamics tools. Some of these tools are as follows:

- Causal loop diagram (CLD)
- Stock and flow maps
- Behavior over time graphs
- Model boundary chart
- Policy structure diagrams

Step 4: Develop the simulation model

At this stage, the modeler needs to identify any changes that may lead to more desirable behavior and the strategies required to face these changes. This stage is identified by three criteria: specification of the structure of the system and the decision rules; determination of the initial conditions and the behavioral relationships; and estimation of the other parameters. The boundaries are considered carefully while developing the model.

Step 5: Test the developed model

The testing step is compulsory for any kind of development. In this step, the developed model can be tested for validity and is compared to a reference model. The robustness and sensitivity of the developed model is also checked at this stage. The feedback about the system can be taken from others and accordingly the model can be modified.

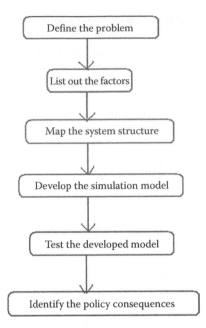

FIGURE 20.8   Steps involved in modeling with system dynamics.

Step 6: Identify the policy consequences
   The actions that must be completed in this step are as follows:
   •   Test all the policies and find the best policy that makes changes.
   •   Identify any environmental constraints that may arise in the system.
   •   Identify any new structure or design issues that may arise and accord-
       ingly modify the model.
   •   Perform a sensitivity analysis for the developed model.
   •   Identify any interaction among the policies and responses.
   •   Organize all these factors to implement the chosen policies.

Figure 20.8 provides an overall diagrammatic representation of the steps involved in the modeling process just discussed.
   To build an effective simulation model, the mapping stage needs the understanding of the system dynamics tools. Thus, Section 20.4 discusses some of the widely used system dynamics tools.

20.4   SYSTEM DYNAMICS TOOLS

System dynamics uses a significant number of tools to represent a mental structure of a behavioral model. These models help to understand the system in a better way. The tools represent logical ways to represent a mental model. There is no particular method for using these tools, but the general guidelines to some of the widely used tools may help users in developing behavioral models with system dynamics tools. Some of these tools are discussed in Sections 20.4.1 through 20.4.3.

20.4.1 Causal Loop Diagrams

CLDs show the relations among the variables in a system. These are mainly used to show the feedback structure of a system. These are simple maps that show the causal links among the variables. The characteristics of CLDs are as follows:

- It is a circular feedback system where a cause may become an effect.
- It identifies the causes of changes in a system and also the key elements that are responsible for the changes.
- It describes how a system actually works since it depicts the causes of the changes. It consists of four symbols:
  - Variables that describe the causes and effects
  - Arrows that indicate the key elements causing the change
  - Symbols associated with arrows that indicate the direction of the influence of the relationships
  - A central symbol that indicates the overall identity of the loop.
- All variables in a CLD must be able to change their values
- It can be either "reinforcing" or "balancing."
- A reinforcing CLD may grow or shrink until a limiting force acts on /hinders it.
- A balancing CLD may oscillate around a particular condition.
- A time delay may be inserted into a CLD if there is a significant time delay between the action of a variable and the reaction of the next variable.

Figure 20.9 shows the notation used in a CLD. Figure 20.10 shows an example of a CLD.

Figure 20.10 shows that if the surface temperature increases, the biological carbon dioxide intake increases. If the biological carbon dioxide intake increases, the atmospheric carbon dioxide decreases, which results in a decrease in the greenhouse effect, which in turn decreases the surface temperature. Thus, this is a negative or balancing loop that balances the surface temperature.

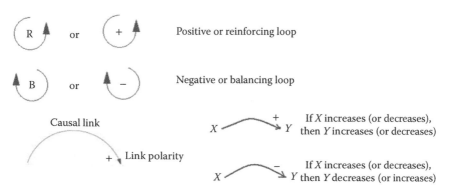

FIGURE 20.9 Symbols used in CLDs.

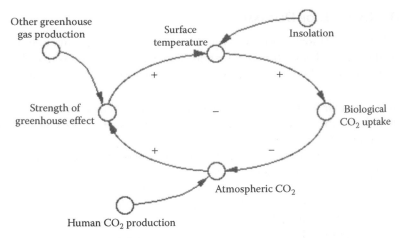

FIGURE 20.10   Example of CLD.

20.4.2 STOCK AND FLOW MAPS

Stock and flow diagrams show the real physical structure of a system. They show the feedback and interdependencies as well as the accumulations and factors that change over time. The characteristics of a stock and flow diagram are given as follows:

- The symbols used by a stock and flow diagram are stock, flow, converter, and connector (Figure 20.11).
- A stock is an accumulator that can increase or decrease over time. The increase or decrease in stock is caused by the flows in the system going through the stock. Stock names are denoted by nouns.
- A flow means transportation. It can be either an inflow or an outflow. An inflow results in an accumulation or increase in a stock, whereas an outflow results in a decrease in a stock.

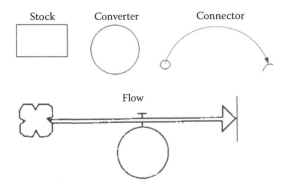

FIGURE 20.11   Symbols of stock and flow diagram.

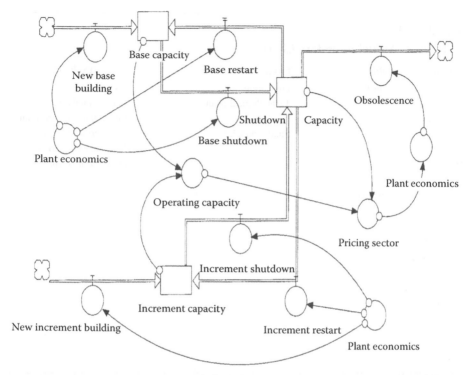

FIGURE 20.12   Example of stock and flow diagram.

- A converter holds information that influences the rate of flow. It may change the value of another converter.
- A connector connects two elements of a system. It takes information from one element and places the information where the connector points to.
- Stock and flow mapping identifies critical stocks and important flows in a system.

An example of stock and flow diagram may make the concept clear. Figure 20.12 shows an example of a stock and flow diagram.

20.4.3   MODEL BOUNDARY CHART

A model boundary chart summarizes the endogenous and exogenous variables that constitute the boundary of the system as a whole. In this way, the chart also depicts the scope of the model. The purpose is to look at the factors external to the system under consideration to observe and analyze their influence over the system.

An example of the model boundary chart [1,2] is given in Table 20.1.

TABLE 20.1

Example of Model Boundary Chart

Endogenous Variables	Exogenous Variables	Excluded
Gross national product	Population	Inventories
Consumption	Technological change	International trade
Investment	Tax rates	Environmental constraints
Savings	Energy policies	Nonenergy resources
Prices		Interfuel substitution
Wages		Distributional equity
Inflation rate		
Labor force participation		
Employment		
Unemployment		
Interest rates		
Money supply		
Debt		
Energy production		
Energy demand		
Energy imports		

20.5 SYSTEM DYNAMICS SOFTWARE

There is a significant number of system dynamics software applications. Both open source software and proprietary software are available for system dynamics. The whole range of software can be classified into five main types: core software, extended software, web-based software, pedagogical software, and documenting software (see Table 20.2). The table also shows examples under each category along with their vendors (within parentheses) for some of the most frequently used system dynamics software.

20.6 CONCLUSION

In this chapter, an overview of the concept of simulation with system dynamics has been presented. System dynamics is basically used for simulating behavioral systems. The modeling steps for developing such simulation models have been presented. Various system dynamics tools and software have also been presented. The overall concept of system dynamics has been introduced so as to show the various aspects of simulation. System dynamics shows only the behavioral aspect among several aspects, which may be an initiative to enhance the pace of research studies in the behavioral areas of study.

TABLE 20.2
Classification of System Dynamics Software

System Dynamics Software Categories	Explanation	Examples and Vendors
Core software	Used to draw system dynamics models only	iThink, STELLA (isee systems), Powersim Studio (Powersim Software), Vensim (Ventana Systems Inc.)
Extended software	Used to build models that include some system dynamics concepts	Anylogic (XJ Technologies), Dynaplan® Smia (Dynaplan), GoldSim (The GoldSim Technology Group), Expose (Attune Group Inc.), Simile (Simulstics)
Web-based software	Used to build web-based system dynamics applications	Forio (Forio Online Simulations)
Pedagogical software	Used for pedagogy or teaching purpose	Sysdea (Strategy Dynamics Limited), InsightMaker (Give Team), NetLogo (Northwestern University)
Documenting software	Used for documenting the developed models	SDM-Doc (Argonne National Labs), Automated Eigenvalue Analysis of System Dynamics models (developed by Ahmed Abdeltawab A.S. Aboughonim, Bahaa E. Aly Abdel Aleem, Mohamed M. Saleh, and Pål I. Davidsen)

REFERENCES

1. Forrester, J. W. (1971). *Principles of Systems*. 2nd Edition. Cambridge, MA: Wright-Allen Press.
2. Sterman, J. D. (2000). *Business Dynamics: Systems Thinking and Modeling for a Complex World*. Singapore: McGraw-Hill.

21 Simulation Software

21.1 INTRODUCTION

Simulation software started its journey in 1955. The period 1955–1960 was the infant period when researchers started to find ways to easily find and analyze the flaws in a system by developing a miniature of the original system. The language FORTRAN was one of the innovations of this period. The main focus of development of various languages at that time was on the unifying concepts and reusable functions.

The period 1961–1965 saw some new simulation software, such as the general purpose simulation system (GPSS). In 1961, IBM launched the GPSS, which was expensive when it was first launched. The GPSS is based on block diagrams and is suitable for queuing models. Another language from the year 1961 is the genometric analysis simulation program (GASP), which was based on ALGOrithmic Language (ALGOL). GASP was a collection of FORTRAN functions. Another extension of ALGOL is SIMULA, which was widely used in Europe. Control and Simulation Language (CSL) was another language of this era. In 1963, the Rand Corporation launched SIMSCRIPT. The largest user of this language was the US Air Force. This language was also influenced by FORTRAN.

The period 1966–1970 is known as the formative period. The existing languages proposed earlier were revised to incorporate new ideas due to the emerging concepts of the time. The number of users also increased, and thus the need for more and more sophisticated features increased. Thus, this period saw newer versions of various user-friendly software such as GPSS SIMSCRIPT II, SIMULA, and ECSL. The new version of SIMULA added the concept of classes and inheritance.

The period 1971–1978 is called the period of expansion. GASP IV and GPSS/H were launched in 1974 and 1977, respectively. The software named SIMULA was also modified. The period 1979–1986 is the period of consolidation and regeneration. Technologies were improved rapidly during this period, and as a result newer software applications were proposed and launched to satisfy the rapidly increasing need of the customers. Minicomputers and personal computers (PCs) came into use during this period. The famous simulation software SLAM II, a derivation of GASP, was proposed and launched. The SIMAN simulation software was also the product of this generation. The modeling of SIMAN is based on block diagrams. SIMAN uses a FORTRAN function for its programming purpose. SIMAN is one of the first major simulation languages for PCs.

The period 1987–2008 is characterized by the incorporation of integrated environments. The rapid growth of PCs and associated technological advancements modified the simulation environment. The simulation software of this period incorporated graphical user interfaces, animation utilities, and data analyzers. Further technological advancement in the period from 2009 to the present is identified by high-tech applications. Thus, the software applications of this period are characterized by virtual reality, improved interfaces, better animation utilities, and agent-based modeling, indicating the inclusion of intelligent components in sophisticated software. Agent-based software applications developed include simulation software such as AnyLogic, Ascape, MASON, NetLogo, StarLogo, Swarm, and RePast.

The high-tech applications of today's technological world has led us to think of almost all concepts in terms of software. In almost all fields of study, people are supposed to learn the respective software to apply the ideas in the respective fields. This fact is especially applicable in the fields of science and technology. Thus, in many cases, researchers are dependent on the software that they apply in their studies. Although the concepts behind the software are required to a great extent to conduct innovative research study, the learning of the current technology is also required to implement new ideas.

In the case of simulation studies, many software applications are observed in the existing literature. The basic definition of simulation indicates the imitation of the actual system under study; such imitation can be done in a variety of ways and needs visualization of the existing system to visualize the problems inherent in the system under study. Visualization of a system is possible by the application of animation or a graphical utility. However, animation is not a compulsory tool in many cases.

Almost all sophisticated software applications provide scope for using graphics of some kind. In addition, we also have standard animation software on hand. However, simulation software not only needs an animation utility but also must have hardcore mathematical and statistical utilities. Since simulation studies consider the uncertainty of all operations, the consideration of statistical probability is a vital part of any simulation software or package. Therefore, to successfully perform a simulation study, in most cases, we need both statistical and mathematical utilities and graphical utilities. Such systems can be developed with many of the modern standard software applications.

However, the existing literature has applied a variety of software applications in almost every field of study. The purpose of this chapter is not just to review the existing literature but also to provide an overview of the various approaches as applied by researchers all over the world. Thus, this chapter is not merely a review. The purpose is to make the reader familiar with the existing approaches as adopted by researchers in various fields of study. Particular emphasis will be given to manufacturing and supply chain areas.

The remaining sections of this chapter are arranged as follows: Section 21.2 provides a brief overview of the types of studies on simulation software in the existing literature. Section 21.3 discusses the various methods for selecting simulation software. Section 21.4 describes the various methods for evaluating simulation software. Section 21.5 provides the conclusion for this chapter.

21.2 TYPES OF STUDIES ON SIMULATION SOFTWARE

A vast number of research studies on simulation software are observed in the existing literature. The area is so appealing to researchers all over the world that the number of research publications in this area increases rapidly each day. A search with the keyword "simulation software" showed 13,372 research papers in the ScienceDirect database; 2,341 research papers in the Taylor & Francis database; 6,685 research papers in the Springer database; and 3,226 research papers in the IEEE database. This clearly indicates the span of applications of the various simulation software applications in various fields of study. Research studies on simulation software can be divided into the following main categories:

- Research studies proposing new simulation software
- Research studies proposing methods of selecting simulation software
- Applications of existing simulation software on various problems
- Research studies proposing methods of evaluating existing simulation software

Research studies proposing new simulation software also include the proposing of small-scale simulation tools. Some simulation tools are even developed on the basis of the context of the experimentation for a particular problem. The number of research studies proposing new simulation software is few in number compared to that of the other research studies on the application of simulation software. Research studies on proposing new simulation software enlighten us with newer software using modern technologies. Most of these software applications are written in languages such as C and C++. However, various other methods are also available:

- Using general purpose languages such as C, C++, and Java
- Using basic simulation languages such as SIMAN, GPSS, SLAM, and SSF
- Using simulation packages such as Arena, SIMUL8, and Enterprise Dynamics

All these languages have some common features, which are listed as follows:

- Statistics collection
- Time management
- Queue management
- Event generation

A search of the topic "selection of simulation software" resulted in 4,568 research papers in the ScienceDirect database, and thus this number indicates the need to find the method of selecting simulation software in various fields of study. Different authors have proposed different methods for selecting simulation software. Research papers in the existing literature on this topic clearly indicate that there is no universal method of selecting simulation software for all fields of study. Particular emphasis and various methods are especially observed in the field of manufacturing. Several authors have endeavored to propose better methods for selecting simulation software. Since there are a significant number of studies in this area, we discuss in Section 21.3 the various methods of selecting such software.

21.3   VARIOUS METHODS OF SELECTING
SIMULATION SOFTWARE

The selection of simulation software depends on both the type of problem under study and the flexibility of the software. Due to these reasons, there is a significant number of research studies on the topic. This section presents some of the approaches of selecting simulation software.

Nikoukaran and Paul [1] reviewed the various methods of selecting simulation software proposed in the literature review. Since practical simulation experimentations are mostly context dependent, the suitability of the simulation software varies considerably, and it is dependent on the problem under study. In addition, the capabilities and the features of the simulation software are also essential factors for selecting simulation software. The selection also varies among the type of organization (education or industry) that adopts any simulation software for its problem.

Nikoukaran and Paul [1] pointed out some of the factors found from the existing literature. These factors are as follows:

- Experience of some other organization
- The features required
- Availability
- Some prespecified tests
- Recommendation by experts

They also listed the steps for selecting simulation software, taken from the existing literature. The steps are as follows:

- Conduct a survey of simulation software used by various other organizations.
- List the suitable software packages available in the market.
- Evaluate and conduct a feasibility study for each of the selected software application.
- Choose the most suitable and feasible software application and perform a trial for some period before purchasing the software.
- Negotiate with the vendor regarding the training and other services that may be required after the software is purchased.
- Purchase the software.

The methods of evaluating various software applications before purchasing a specific one will be discussed in Section 21.4. However, the existing literature provides a variety of criteria or features on the basis of which the selection can be made. The list of such criteria is exhaustive since the authors have found these criteria based on various problems and needs. Most of the authors have categorized all the criteria into some groups based on some features. Some of these features for classifying the set of criteria are as follows:

- Interface between the software and the user
- Debugging capability of the software
- Interactive features

- Documentation system of the software
- Troubleshooting capabilities
- Database storage capabilities

In addition, there are other groups as proposed by various authors. For example, Law and Kelton [2] have classified simulation software on the basis of the following characteristics:

- General features
- Animation
- Statistical capabilities
- Customer support
- Analysis of output from simulation experiment

Banks et al. [3] have identified the following defining characteristics:

- Robust features
- Quantitative factors
- Basic features
- Special features
- Cost

Holder [4] has identified the defining characteristics:

- Technical features
- Needs of the users inside the organization
- Customer needs
- Functionality
- Commercial features
- Development features

Mackulak et al. [5] have identified the following characteristics:

- General features
- Features related to data acquisition
- Method of developing models
- Verification and validation utilities
- Analysis of output from simulation experiment
- Documentation facility
- User interface
- Model execution

In this way, numerous other classification criteria have been proposed by various other authors. A gist of such factors is listed in Table 21.1.

Another approach proposed by Hlupic and Paul [6] is a phased method for software selection. This method also considers discrete or continuous simulation,

TABLE 21.1
List of Features for Classifying Selection Criteria

General Features	Animation	Statistical Capabilities	Customer Support	Output Analysis	Documentation	Input Features
Technical features	Functionality	Commercial features	Cost	Output reports	Data acquisition	Simulation experiment
Basic constructs	Special constructs	User interface	Verification	Validation	Training	Ease of use
Statistical facilities	Printed manuals	Hardware	Installation	User interaction	Model development	Efficiency
Testing	Debugging capability	Integration with other systems	Processing features	Model specification	Modeling assistance	Vendor
Coding aspects	Model execution	Simulation project data				

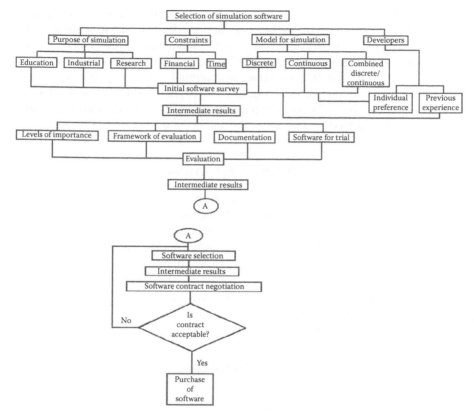

FIGURE 21.1 Method of software selection. (Adapted from Hlupic, V. and Paul, R. J., *Computer Integrated Manufacturing Systems*, 9, 49–55, 1996.)

education, and other factors (Figure 21.1). In this method, at first, the software is selected on the basis of the purpose for the simulation study, the constraints within which the simulation study is to be performed, the model that is to be simulated, and the types of developers. The purpose of simulation depends on the application of the simulation experiment. Thus, the simulation study may be conducted for educational, industrial, or research purposes.

There can be several constraints under which the simulation experiment is conducted. The most prominent ones are financial and time constraints. Moreover, simulation model to be constructed may be a discrete simulation model, a continuous simulation model, or a combination of both types of models. The developer may use individual preferences while developing the model or may use any previous experience for development. Based on all these factors, an initial survey may be conducted. This survey may be based on some predefined questionnaire that provides valuable information and guidance to the organization or individual who is planning to purchase the simulation software.

From the results obtained from the survey, decisions may be made on choosing the importance level, the framework for software evaluation, the components related to documentation, and the software for trial. This means that the software applications now may be evaluated on the basis of various criteria to see which is feasible for purchase. Several alternative software applications may be evaluated before an actual choice is made. The next step after selecting the software is the signing of the contract for the software. If the contract is acceptable for both parties involved, the software is finally purchased. From these steps, it is clear that the evaluation of the available existing software is an important step in software selection. Section 21.4 discusses the method of software evaluation in detail.

21.4   SIMULATION SOFTWARE EVALUATION

There are a number of research articles in the existing literature on the evaluation of simulation software. However, this chapter is mainly based on the study of Nikoukaran et al. [7]. This study presents a hierarchical framework for describing the various factors for evaluating simulation software. The entire set of software evaluation criteria has been divided into three major divisions:

1. Criteria related to vendors of simulation software
2. Criteria based on the software
3. Criteria based on the users' preferences

The related criteria for vendors are as follows:

- The vendor's background
- The documentation provided by the vendor
- The other associated services delivered by the vendor
- The service before purchasing the software

The vendor's background involves the history and background of the vendor, the type of quality of the software delivered, and the reputation of the other products delivered by the vendor. The relevant criteria for software aspects are the speed, the accuracy of the results, the age of the software, and the reliable references for purchasing the software.

The documentation factors are as follows:

- Whether the manual for the software purchased is provided
- Whether tutorials are provided for future reference and learning of the software
- Whether sufficient examples are provided for the application of the simulation software
- Whether sufficient troubleshooting options are available and are documented properly for possible future requirements
- Whether the statistical background for the software is provided in the documentation
- Whether a proper index is provided

The other associated services are the postservices and promised services from the vendor. These services include the following:

- Training provided
- The additional cost of training
- Periodic maintenance provided by the vendor
- Technical support provided by the vendor
- The potential benefit of the consultancy provided by the vendor
- Provision of getting updates for newer versions of the software to be purchased
- Contact facilities provided by the vendor, such as useful phone numbers and online contacting facilities
- Promptness of the service when required

The prepurchase issues include the following:

- On-site demonstration
- Free trial for the software
- Provision of demo disks

The major groups of criteria related to the software aspects are as follows:

- Model input
- Model execution
- Animation facilities
- Testing and debugging
- Output of the simulation experiment

21.4.1 MODEL INPUT

There are a number of criteria related to model input, some of which are as follows:

- Development of modules that can be reused
- Various tools used for building a module
- Provisions for merging modules if required
- Conditional branching
- Various aspects related to coding
- Statistical distributions
- Queuing policies
- Aspects related to inputs to simulation experiment

Some of the important coding aspects are as follows:

- *Availability of source code for the simulation language.* This is especially important for those researchers who endeavor to embed some other package with a simulation software application or require significant coding involving the internal function of the simulation software application.

- *Programming tools and compilation speed.* The compilation speed needs to be high to speed up the execution of the simulation experiment. And programming tools differ with the software. Thus, the flexibility of a software application depends on its programming tools.
- *Provision of linking or embedding other languages or packages.* This is very important for many innovative research studies. The need of the problem may lead to the need to connect simulation software with the other software. For example, most of the simulation software applications are compatible with the MS Excel package. Because of this facility, various outputs can be shown in the desired way by the researcher. However, every software application seems to have limitations to some extent regarding this particular feature.
- *Availability and variety of the built-in functions.* These influence the flexibility of simulation or any other type of software. The user-defined functions are also dependent upon the built-in or system functions of the software. Thus, the capabilities of built-in functions are an important feature of simulation software.
- *Global values and their number.* They can result in more flexible programs. In addition, sometimes the portability and reusability of a program also increases because of the use of global variables.

21.4.2 MODEL EXECUTION

The criteria for model execution are related to various options of simulation experimentation. They are as follows:

- Number of runs, which is decided by the user
- Batch runs
- Warm-up period
- Initial condition
- Initialization of variables
- Provision for parallel execution
- Speed of execution of simulation experiment
- Executable models

21.4.3 ANIMATION FACILITIES

Another important aspect of simulation in today's high-tech age is the animation aspect of simulation software. Animation is used in simulation to realize the actual simulated scenario. Animation movement can show movements of various parts in the system. Various icons are used to represent various parts or components of a system under simulation. Another important aspect is the running of the animated model. Various settings to run a simulation model are adjusted before running the animated model. Thus, the running conditions, the look, or the layout of the simulation screen must be considered for a successful animation model.

21.4.4 Testing and Debugging

Another set of criteria can be defined for testing and debugging. Some of the important criteria are as follows:

- Validation and verification of the output or results obtained from the simulation experiment
- Level of interaction during testing and debugging, which is required for some cases where the output is dependent on the user's opinion
- Snapshots
- Simulation clock or time-related options of simulation experimentation
- Provision of setting breakpoints to check and monitor the execution of the simulation experiment
- Various display features
- Providing step functions and tracing for monitoring purposes
- Other debugging facilities

21.4.5 Output of the Simulation Experiment

The last group of criteria related to the simulation experiment is the set of criteria for the output. Some of the criteria for this category or division are given as follows:

- Report-making options for the output generated—standard or customized report making
- Database generated from the output—this database generally contains various types of outputs.
- Provision for graphical presentation of the output generated, such as a pie chart, bar graph, line diagram, or histogram
- Provision for providing the output in a file, or printing the output, and exporting the output to some other application for further use of the results or for better presentation purposes
- Provision for performing various analyses of the results obtained, such as finding correlations among different variables, finding confidence intervals, hypothesis testing, optimization of the output generated, regression analysis of the results obtained, experimental designs, and time series analysis
- The last category is related to the aspects of the users. This group indicates how the users use the simulation. For example, the user may opt for either a discrete or continuous type of simulation or a combination of both types. The experience of the user in simulation or other types of software is another important criterion, since it determines the various user settings that in turn shape the success of a simulation experiment. In addition, there are various other factors that can shape the success of the simulation experimentation, such as the field of study in which the simulation study is conducted, the operating system used, the price of the simulation software, maintenance, training, and the discount offered.

21.5   CONCLUSION

In this chapter, various aspects of simulation software have been discussed. The data presented in each of the sections are taken from the existing literature. The existing literature mainly focuses on the application of various simulation software applications, method of selecting simulation software, and evaluating the simulation software. Since this chapter is not intended to be a review, the various applications of simulation software in all fields of study have not been discussed here. In addition, since several other chapters have listed and discussed the various simulation software, the list of software applications in various fields of study has not been provided in this chapter. The overall purpose of this chapter is to guide the readers to the various aspects of selecting and evaluating the available simulation software before purchasing or using them. Thus, this chapter has mentioned and discussed several selection and evaluation criteria for purchase of simulation software. Various types of simulation software have also been introduced in Section 21.2. Thus, this chapter is expected to assist readers in knowing the above-mentioned aspects of simulation software.

REFERENCES

1. Nikoukaran, J. and Paul, R. J. (1999). Software selection for simulation in manufacturing: A review. *Simulation Practice and Theory* 7(1): 1–14.
2. Law, A. M. and Kelton, W. D. (1991). *Simulation Modeling and Analysis*. 2nd Edition. Singapore: McGraw-Hill.
3. Banks, J. (1991). Selecting simulation software. *Proceedings of the 1991 Winter Simulation Conference*, December 8–11, Phoenix, AZ. New York: ACM Press, pp. 15–20.
4. Holder, K. (1990). Selecting simulation software. *OR Insight* 3(4): 19–24.
5. Mackulak, G. T., Savory, P. A., and Cochran, J. K. (1994). Ascending important features for industrial simulation environments. *Simulation* 63: 211–221.
6. Hlupic, V. and Paul, R. J. (1996). Methodological approach to manufacturing simulation software selection. *Computer Integrated Manufacturing Systems* 9(1): 49–55.
7. Nikoukaran, J., Hlupic, V., and Paul, R. J. (1999). A hierarchical framework for evaluating simulation software. *Simulation Practice and Theory* 7: 219–231.

22 Future Trends in Simulation

22.1 INTRODUCTION

Simulation is an extensive field of study. The field is rich with wide applications in almost all fields of science and technology. The existing literature shows that even a small program written in a simple language can be thought of as an endeavor in simulation if the program models some sort of system in some way. Thus, even a simple language such as C can be used as a simulation language. Basically, C was once used as a very effective language to simulate small-scale systems. However, with the advancement of technology, the need for more user-friendly and sophisticated languages increased and newer and more advanced software came into our hand.

C is a procedural language that can be used to develop the programming sense in a programmer's mind. However, with the increase in complexity of the application in various languages, the need for easy-to-use software increased so that each step of the logic of a program does not have to be written in code. Thus, we came across the concept of object-oriented programming languages such as C++. These object-oriented languages were widely used especially in developing various intelligent applications, such as agent-based systems. The world of simulation was also greatly influenced by this advancement. Nowadays, researchers are using virtual systems to simulate real-world applications. Animation and virtual reality play a vital role in simulating various complex systems to understand and modify them. These technologies are further advancing along with the hardware. The basic personal computer (PC) today has a configuration that was not generally available even 5 years ago. Such high-configuration PCs are suitable for virtual computing, which is a very recent trend.

The latest Winter Simulation Conference (WSC), the prime conference on simulation, showed various thoughts on simulation in the various fields of study as presented. The WSC in 2012 showed the endeavor of automatic object-oriented code generation for simulation practice, advanced database-driven architectures, research studies on incremental and flexible agent-based modeling, application of simulation in pipeline networks, simulation modeling for supply-chain operations reference (SCOR) models, interaction of visual simulation, and new efforts in the area of simulation optimization. Thus, these can be taken as the latest trends of simulation. In addition, the literature shows an increasing enthusiasm in the field of cloud computing applications. Thus, this chapter introduces some of the promising trends in the field of simulation.

Section 22.2 introduces various .NET technologies that can be used in simulation. Section 22.3 introduces the concept of cloud virtualization. Section 22.4 shows directions to future research studies in the field of simulation. Section 22.4 provides the conclusion for this chapter.

22.2   .NET TECHNOLOGIES

.NET technology is one of the most widely used software technologies in use today. Most industrial applications today are based on .NET technologies [1]. The main languages that can be used in the programming for simulation include C# .NET and Visual Basic .NET. First we give a brief introduction to .NET.

Microsoft .NET offers an object-oriented view of the Windows operating system and includes hundreds of classes that encapsulate all the most important Windows kernel objects. For example, a portion of .NET named GDI+ contains all the objects that create graphic output in a window. Depending on the specific Windows version, in the future, these objects might use Windows graphics device interface (GDI) functions, the DirectX library, or even OpenGL, but as far as the developer is concerned, there is only one programming model to code against.

Microsoft provides several languages with .NET, including Visual Basic .NET, C#, Managed C++, Visual J#, and JScript. Many C++ developers have switched or are expected to switch to C#, which is considered the best language for .NET applications. C# is actually a great language that takes the best ideas from other languages, such as C++ and Java. But the truth is that C# and Visual Basic .NET are roughly equivalent, and both of them let you tap the full power of the .NET Framework. The two languages have the same potential, so you are free to choose the one that makes you more productive. Execution speed is also equivalent for all practical purposes, because the C# and Visual Basic compilers generate more or less the same code. ASP.NET is arguably the most important portion of the .NET Framework. ASP.NET comprises two distinct but tightly related technologies: web forms and extensible markup language (XML) web services. Web forms are used for Internet applications with a user interface and are meant to replace application service provider (ASP) applications, although you can still run ASP and ASP.NET on the same computer. Web services are for Internet applications without a user interface.

The best way to understand how .NET works is to look at the many layers in the .NET Framework, shown in Figure 22.1. Each of these components is described

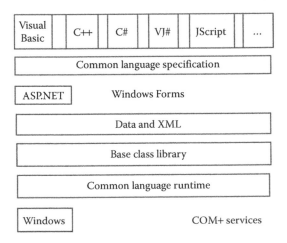

FIGURE 22.1   Layers of .NET Framework.

in the following. At the bottom of the hierarchy sits the Windows application programming interface (API). The .NET Framework offers an object-oriented view of the operating system's functions but does not replace them, so you should not forget that most calls into the .NET Framework are ultimately resolved as calls into one of the Windows kernel dynamic-link libraries (DLLs). The common language runtime is the first layer that belongs to the .NET Framework. This layer is responsible for .NET base services, such as memory management, garbage collection, structured exception handling, and multithreading. If .NET is ever to be ported to non-Windows architectures—as of this writing, a few projects are pursuing this goal—writing a version of the common language runtime for the new host must be the first step. The runtime is contained in the mscoree.dll file, and all .NET applications call a function in this DLL when they begin executing.

The base class library (BCL) is the portion of the .NET Framework that defines all the basic data types, such as System.Object (the root of the .NET object hierarchy), numeric and date types, string types, arrays, and collections. The BCL also contains classes for managing .NET core features, such as file I/O, threading, serialization, and security. The way types are implemented in the BCL follows the common type system (CTS) specifications. For example, these specifications dictate how a .NET type exposes fields, properties, methods, and events. They also define how a type can inherit from another type and possibly override its members. Because all .NET languages recognize these specifications, they can exchange data, call into each other's classes, and even inherit from classes written in other languages. The BCL is contained in the mscorlib.dll component.

The data and XML layer contains the .NET classes that work with databases and XML. Here you can see that the support for XML is built right into the .NET Framework, rather than accessing it through external components, as is the case with pre-.NET languages. In fact, XML can be considered as the format that .NET uses to store virtually any kind of information. All the .NET configuration files are based on XML, and any object can be saved to XML with just a few statements. Of course, the .NET Framework comes with a powerful and fast XML parser.

The next two layers are ASP.NET and Windows Forms, which are located at the same level in the diagram. These portions of the framework contain all the classes that can generate the user interface—in a browser in the former case and using standard Win32 Windows in the latter case. As explained earlier, ASP.NET comprises both web forms and web services. Even though these two portions appear at the same level in the diagram, and in spite of their similarities, these technologies are very different. Web forms run on the server and produce hypertext markup language (HTML) code that is rendered in a browser on the client (which can run on virtually any operating system), whereas Windows Forms run on the client (and this client must be a Windows machine). However, you can mix them in the same application, to some extent at least. For example, you can have a .NET application that queries a remote web service through the Internet and displays its result using Windows Forms.

The common language specification (CLS) is a set of specifications that Microsoft has supplied to help compiler vendors. These specifications dictate the minimum group of features that a .NET language must have, such as support for signed integers of 16, 32, or 64 bits; zero-bound arrays; and structured exception handling. A compiler vendor is free to create a language that exceeds these specifications—for example, with unsigned integers or arrays whose lowest index is nonzero—but a well-designed application should never rely on these non-CLS-compliant features when communicating with other .NET components because the other components might not recognize them. Interestingly, Visual Basic .NET matches the CLS specifications almost perfectly, so you do not have to worry about using non-CLS-compliant features in your applications. C# developers can use a Visual Studio .NET option to ensure that only CLS-compliant features are exposed to the outside.

At the top of the diagram in Figure 22.1 are the programming languages that comply with CLS. Microsoft offers the following languages: Visual Basic .NET, C#, Managed C++, Visual J#, and JScript. Many other languages have been released by other vendors or are under development, including Perl, COBOL, Smalltalk, Eiffel, Python, Pascal, and APL.

All .NET languages produce managed code, which is a code that runs under the control of the runtime. Managed code is quite different from the native code produced by traditional compilers, which is now referred to as unmanaged code. Of all the new language offerings from Microsoft, only C++ can produce both managed and unmanaged codes, but even C++ developers should resort to unmanaged code only if strictly necessary—for example, for doing low-level operations or for performance reasons—because only managed code gets all the benefits of the .NET platform. The similarity among all languages has three important consequences. They are as follows:

1. The execution speed of all languages tends to be the same.
2. Language interoperability is ensured because all the languages use the same data types and the same way to report errors, so you can write different portions of an application with different languages without worrying about their integration.
3. Learning a new .NET language is surprisingly effortless if you have already mastered another .NET language.

Section 22.2.1 gives an overview of C# .NET.

22.2.1   C# .NET

In cooperation with the .NET Common Language Runtime, C# provides a language to use for component-oriented software, without forcing programmers to abandon their investment in C, C++, or COM code. C# is designed for building robust and durable components to handle real-world situations.

It is a "component-centric" language, in that all objects are written as components, and the component is the center of the action. Component concepts, such as properties, methods, and events, are the first-class citizens of the language and the underlying runtime environment. Declarative information (known as attributes) can be applied to components to convey design-time and runtime information about the component to other parts of the system. Documentation can be written inside the component and exported to XML. C# objects do not require the creation or use of header files, interface definition language (IDL) files, or type libraries. Components created by C# are fully self-describing and can be used without a registration process.

C# provides an environment that is simple, safe, and straightforward. Error handling is not an afterthought, with exception handling being present throughout the environment. The language is typesafe, and it protects against the use of variables that have not been initialized, unsafe casts, and other common programming errors. C# provides the benefits of an elegant and unified environment, while still providing access to "less reputable" features—such as pointers—when those features are needed to get the job done.

C# is built on a C++ heritage and should be immediately comfortable for C++ programmers. The language provides a short learning curve, increased productivity, and no unnecessary sacrifices. Finally, C# capitalizes on the power of the .NET Common Language Runtime, which provides an extensive library support for general programming tasks and application-specific tasks. The .NET Runtime, Frameworks, and languages are all tied together by the Visual Studio environment, providing one-stop shopping for the .NET programmer.

C# is an object-oriented language, and thus all its constructs are objects in nature. An object is merely a collection of related information and functionality. An object can be something that has a corresponding real-world manifestation (e.g., employee), something that has some virtual meaning (e.g., a window on the screen), or something that has just some convenient abstraction within a program (e.g., a list of work to be done). An object is composed of the data that describe the object and the operations that can be performed on the object. Information stored in an object, for example, might be identification information (name, address) or work information (job title, salary). The operations performed might include creating an employee paycheck or promoting an employee.

The characteristics of an object-oriented language are as follows:

- Inheritance, for example, the model car Maruti inherits the compulsory property of car as a whole.
- Polymorphism means having different forms for the same concept.
- Virtual function
- Data abstraction
- Encapsulation means that the data and the function are bundled together in a compact way.
- Visibility indicates that the functions of the objects are not accessible easily. This feature also indicates the security of the data and function contained in an object.

All .NET languages use namespace. Namespaces in the .NET Runtime are used to organize classes and other types into a single hierarchical structure. The proper use of namespaces will make classes easy to use and prevent collisions with classes written by other authors.

Namespaces can also be thought of as a way to specify really long names for classes and other types without having to always type a full name.

However, the application of C# can be effective in case simulation studies. Bandyopadhyay and Bhattacharya [2] used C# .NET to simulate a routing scenario to implement a particular routing strategy. The routing strategy was based on the mating behavior of a species of spider known as Tarantula, in which the female spider sometimes eats the male spider just after mating for the want of food or for genetic purposes. The routing proposed in the research study opted to choose the immediate neighbor of a node instead of establishing the end-to-end path from the source to the destination. To serve the purpose, several criteria were evaluated and the most promising neighbor was chosen as the next neighbor. An agent-based framework was used which was implemented by using C# .NET language. In the agent-based framework discussed in the works of Bandyopadhyay and Bhattacharya [2], there are two types of agents—master agent and several worker agents. Each of the worker agents performed some tasks to evaluate the various aspects of the neighbor. For example, the shortest path agent found the shortest path from each of the immediate neighbors to the destination; the deadlock agent checked whether there was any chance of getting blocked by deadlock for any of the immediate neighbors; and so on. These results from the worker agents were input to the master agent. After delivering all the required data to the master agent, the worker agents were killed to save valuable computational resources. This was an example of network simulation where a particular problem was simulated by using C# .NET. Thus, C# .NET can be used for simulation in various disciplines.

22.3 CLOUD VIRTUALIZATION

Virtualization, in its broadest sense, is the emulation of one of more workstations/ servers within a single physical computer. Put simply, virtualization is the emulation of hardware within a software platform. This allows a single computer to take on the role of multiple computers. This type of virtualization is often referred to as full virtualization, allowing one physical computer to share its resources across a multitude of environments. This means that a single computer can essentially take the role of multiple computers. However, virtualization is not limited to the simulation of entire machines. There are many different types of virtualization, each for varying purposes. One of these is used by almost all modern machines today and is referred to as virtual memory. Although the physical locations of the data may be scattered across a computer's random access memory (RAM) and hard drive, the process of virtual memory makes it appear such that the data are stored contiguously and in order. The redundant array of independent disks (RAID) is also a form of virtualization along with disk partitioning, processor virtualization, and many other virtualization techniques.

Virtualization allows the simulation of hardware via software. For this to occur, some type of virtualization software is required on a physical machine. The most well-known virtualization software in use today is VMware. VMware will simulate the hardware resources of an ×86-based computer to create a fully functional virtual machine. An operating system and associated applications can then be installed on this virtual machine, just as would be done on a physical machine.

Multiple virtual machines can be installed on a single physical machine as separate entities. This eliminates any interference between the machines, each operating separately.

The four main objectives of virtualization, demonstrating the value offered to organizations, are as follows:

1. Increased use of hardware resources
2. Reduced management and resource costs
3. Improved business flexibility
4. Improved security and reduced downtime

However, no one could have imagined the massive growth in the use of computer technology, and this created new IT infrastructure demands as well as problems. Some of these problems are as follows:

- Low hardware infrastructure utilization
- Rising physical infrastructure costs
- Rising IT management costs
- Insufficient disaster protection
- High maintenance and costly end-user desktops

However, the benefits of virtualized technologies are as follows:

- Easy to monitor the problems inherent in the system
- Eliminates compatibility issues
- Any kind of error within a virtual machine will not affect any other virtual machine.
- Increased security of data inside the virtual machines
- Resources of a machine are used efficiently.
- A virtual machine can easily be transferred from one machine to another machine without any changes to functionality.
- One or more virtual machines can be set up as test machines without hampering the day-to-day business activities.
- The hard drive of a virtual machine stays on the physical machine as a single drive, and therefore the hard drive of a virtual machine can easily be replicated to another machine or another drive in the same machine.
- The cost of hardware is reduced significantly.
- Services that can be in conflict with one another can be installed in different virtual machines even on the same computer. For example, a Windows operating system and a Linux operating system can be installed separately in two virtual machines.
- Easier to maintain

However, besides the entire computer, many other resources can be virtualized. For example, disk virtualization is one of the oldest forms of storage virtualization. The physical form of a magnetic disk is a compilation of cylinders, heads, and sectors.

Each disk is different based on the number of cylinders, heads, and sectors, which changes the capacity of the disk. To read or write using the disk, some form of addressing is required. To complete this addressing, the exact physical property of every magnetic disk would have to be known—an impossibility. Virtualization is completed by the disks' firmware to transform the addresses into logical blocks for use by applications and operating systems, called logical block addressing (LBA). As magnetic disks are used, some blocks may not be able to store and retrieve data reliably. Disk virtualization allows the magnetic disk to appear defect-free, releasing the operating system to focus on other processes.

File systems virtualization provides file systems to multiple clients regardless of the operating systems on those clients. Using a networked remote file system such as Network File System (NFS) or Common Internet File System (CIFS), this abstraction demonstrates the most important feature of storage virtualization: location transparency. Another form of file system virtualization assists in database management. Most data for databases are located on raw disk drives to maximize performance but are cumbersome to manage. Putting these data into a file system makes the data easier to manage but causes problems in performance. Database virtualization combines the best of both worlds. Figure 22.2 shows the various forms of cloud virtualization.

Some of the service management processes involved in virtualization include the following:

- Demand management
- Capacity management
- Financial management
- Availability management
- Information security management
- IT service continuity management
- Release and deployment management
- Service asset and configuration management

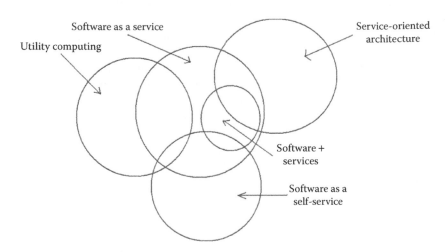

FIGURE 22.2   Various forms of cloud virtualization.

- Knowledge management
- Incident management
- Problem management
- Change management
- Service desk function

Cloud computing [3] refers to the processing and storage of data through the Internet. Computing and storage become "services" rather than physical resources. Files and other data can be stored in the cloud and can be accessed from any Internet connection. It is a style of computing where IT-related capabilities are provided "as a service," allowing users to access technology-enabled services from the Internet, or cloud, without knowledge of, expertise with, or control over the technology infrastructure that supports them.

Software as a service (SaaS) is a form of computing that is closely related to the cloud. As with cloud computing, SaaS customers tap into computing resources offsite that are hosted by another company. However, the difference lies in scale. Cloud computing platforms may combine thousands of computers and storage networks for backup, while some SaaS applications may operate via a connection between only two or three computers. This means that most SaaS solutions fall under the larger cloud computing definition. However, there are a few distinguishing features of SaaS that separate these solutions from cloud computing. SaaS is a software application that is owned, delivered, and managed remotely by one or more providers. It also allows sharing of application processing and storage resources in a one-to-many environment on a pay-for-use basis, or as a subscription. Many cloud computing offerings have adopted the utility computing model. Utility computing can be analogized to traditional computing like the consummation of electricity or gas. IT computing becomes a utility, similar to power, and organizations are billed either based on usage of resources or through subscription.

Cloud computing promises to increase the velocity with which applications are deployed, increase innovation, and lower costs, all the while increasing business agility. Sun takes an inclusive view of cloud computing that allows it to support every facet, including the server, storage, network, and virtualization technology that drives cloud computing environments to the software that runs in virtual appliances that can be used to assemble applications in minimal time. This white paper discusses how cloud computing transforms the way we design, build, and deliver applications, and the architectural considerations that enterprises must make when adopting and using cloud computing technology.

Cloud computing extends this trend through automation. Instead of negotiating with an IT organization for resources on which to deploy an application, a compute cloud is a self-service proposition where a credit card can be used to purchase compute cycles, and a web interface or API is used to create virtual machines and establish network relationships between them. Instead of using a long-term contract for services with an IT organization or a service provider, one may use compute clouds, which work on a pay-by-use or pay-by-the-sip model where an application may exist to run a job for a few minutes or hours, or it may exist to provide services to customers on a long-term basis. Compute clouds are built as if applications are temporary, and billing is based on resource consumption: CPU hours used, volumes of data moved, or gigabytes of data stored.

Cloud computing concepts can be used to simulate various manufacturing supply chain scenarios. Such simulation can be very effective as well as very inexpensive.

22.4 CONCLUSION

In this chapter, a brief conclusion about the future trends of simulation has been presented. As reference, the authors have followed the WSC proceedings for some years and exerted their effort to find the trend in this area. Recent research topics in various simulation journals were also consulted. The authors have identified two broad areas where a very small number of research works have been conducted but the areas are very promising. Because of the technological trends, it is expected that these fields are going to be preferred as effective simulation aids in the near future. Thus, this chapter has given a brief description of .NET technologies and virtualization using cloud computing.

REFERENCES

1. Stephens, R. (2011). *Start Here! Fundamentals of Microsoft .NET Programming.* Sebastopol, CA: Microsoft Press.
2. Bandyopadhyay, S. and Bhattacharya, R. (2013). Finding optimum neighbor for routing based on multi-criteria, multi-agent and fuzzy approach. *Journal of Intelligent Manufacturing*, doi:10.1007/s10845-013-0758-6.
3. Sarna, D. E. Y. (2011). *Implementing and Developing Cloud Computing Applications.* Boca Raton, FL: CRC Press.

Index

Note: Locators followed by "*f*" and "*t*" denote as figures and tables in the text

Printed and bound by CPI Group (UK) Ltd, Croydon, CR0 4YY

18/10/2024

01776259-0011